Cambridge Studies in Ecology presents balanced, comprehensive, up-to-date, and critical reviews of selected topics within ecology, both botanical and zoological. The Series is aimed at advanced final-year undergraduates, graduate students, researchers, and university teachers, as well as ecologists in industry and government research.

It encompasses a wide range of approaches and spatial, temporal, and taxonomic scales in ecology, including quantitative, theoretical, population, community, ecosystem, historical, experimental, behavioral and evolutionary studies. The emphasis throughout is on ecology related to the real world of plants and animals in the field rather than on purely theoretical abstractions and mathematical models. Some books in the Series attempt to challenge existing ecological paradigms and present new concepts, empirical or theoretical models, and testable hypotheses. Others attempt to explore new approaches and present syntheses on topics of considerable importance ecologically which cut across the conventional but artificial boundaries within the science of ecology.

Spiders in ecological webs

CAMBRIDGE STUDIES IN ECOLOGY

Editors
H. J. B. Birks *Botanical Institute, University of Bergen, Norway*
J. A. Wiens *Department of Biology, Colorado State University, USA*

Advisory Editorial Board
P. Adam *School of Biological Science, University of New South Wales, Australia*
R. T. Paine *Department of Zoology, University of Washington, Seattle, USA*
F. I. Woodward *Department of Animal & Plant Sciences, University of Sheffield, UK*

Spiders in ecological webs

DAVID H. WISE

CAMBRIDGE
UNIVERSITY PRESS

Published by the Press Syndicate of the University of Cambridge
The Pitt Building, Trumpington Street, Cambridge CB2 1RP
40 West 20th Street, New York, NY 10011–4211, USA
10 Stamford Road, Oakleigh, Melbourne 3166, Australia

First published 1993
First paperback edition 1995

Printed in Great Britain at the University Press, Cambridge

A catalogue record for this book is available from the British Library

Library of Congress cataloguing in publication data

Wise, David H.
Spiders in ecological webs/David H. Wise.
  p.   cm. – (Cambridge studies in ecology)
Includes bibliographical references (p.   ) and index.
ISBN 0 521 32547 1.
1. Spiders – Ecology.   2. Spider populations.   3. Spiders –
Ecophysiology.   4. Food chains (Ecology) I. Title. II. Series.
QL458.4.W57   1993
595.4′405 – dc20   92-11137   CIP

ISBN 0 521 32547 1 hardback
ISBN 0 521 31061 X paperback

Contents

Preface

Ecologists love to spin yarns. Some are wonderfully intricate tales that radiate connections spanning voids separating far-flung supports. Such spinning work dazzles but often is short lived, as is the orb web that so carefully crafted succumbs eventually to wind, rain and dew. The patient spider, not unlike the ecologist, ingests the remnants of its artwork and spins again, sometimes in another site or with a different orientation.

Others try different strategies. They pick a smaller space and spin an irregular maze that lasts longer but requires continual additions and refinements. Yet another strategy combines the two, like that of the labyrinth spider, which spins an orb to capture unwary prey but hides from enemies in an adjoining, irregular framework. And then there are the thieves, the kleptoparasites, who do not build their own webs but instead use the snares of others to make a living. Or consider the wandering spiders, webless marauders who wait in ambush or stealthily stalk their prey, who retrace the past with draglines, safety lines clutched frantically to halt the plunge to unknown depths after a wrong turn has taken them over the edge.

Analogies between ecologists and spiders become tenuous if stretched too far. The spider, for example, never becomes trapped in its own web. Neither do ecologists if they're careful. In the chapters that follow I will evaluate generalizations woven by ecologists to explain how spiders connect with other organisms, how spiders fit into their ecological webs. It's fair to ask what criteria I will use to evaluate these generalizations. As much as possible, I will rely upon their agreement with the results of controlled field experiments, an approach that has been a bias since my earliest days as an ecologist. In fact, this approach to ecology has been intertwined with my interest in spiders since my first field manipulation as a graduate student.

My fascination with spiders began with the discovery that I could establish replicated populations of a sheet-web spider at different densities on small, unenclosed bushes, and that I could simultaneously supplement natural prey levels. Thus in one experimental design I was able to manipulate the density of a consumer and the rate of supply of its resource in an open, natural system. Hypotheses concerning resource limitation and competition could thus be tested directly for an important terrestrial carnivore. Since these early experiments I have come to realize that not all spider populations can be manipulated so easily, yet my fascination with them increases as I learn more about their diversity and their complicated connections within the ecological web.

Spiders continue to capture scientific prey, from sociobiologists to ecosystem ecologists. Many victims, perhaps even an occasional vertebrate biologist, have started research with spiders because of their promise as experimental organisms, and continue because spiders are too fascinating to abandon. A major motivation for writing this book is to encourage this trend by introducing spiders to an even wider ecological audience.

Acknowledgments

It is self-evident but nonetheless deserving of acknowledgment, that this book is a direct outgrowth of the accomplishments of other ecologists. I have profited greatly from reviewing their research, and hope that my appreciation is apparent in the attention I have given their studies. Some comments are laudatory, some may appear neutral, others are critical, but all are made in the spirit of constructive criticism and synthesis.

Several persons have provided additional information on their published research or have furnished unpublished data, manuscripts, or dissertations. Most of this input was given while I was writing the book, but some came in correspondence that preceded the start of the project. In particular I thank the following, whose contributions appear in the following pages: Richard Bradley, Hartmut Döbel, Allan Markezich, John Martyniuk, Douglass Morse, Gary Polis, Susan Riechert, Matthias Schaefer, David Spiller, Søren Toft and George Uetz.

I have benefited immensely from helpful comments on drafts of the manuscript. Paul Reillo, Susan Riechert, Matthias Schaefer, Tom Schoener, David Spiller and James Wagner each generously agreed to read many of the chapters; they all provided very thoughtful critiques. I also received valuable feedback from other colleagues who carefully read and commented upon a particular chapter: Hartmut Döbel, Matthew Greenstone, Nancy Kreiter and George Uetz.

My father, Gilbert Wise, drew the spider illustrations that appear throughout the book. I thank him not only for this contribution, but also for his life-long fascination with natural history. My father never explicitly encouraged me to become an ecologist, but parents do have a way of influencing their children, even if unintentionally.

1 · The spider in the ecological play

Setting the stage

G. Evelyn Hutchinson (1965) succinctly described the world view of most ecologists when he titled a collection of essays 'The Ecological Theater and the Evolutionary Play.' Ecologists routinely postulate roles for their favorite organisms in the ecological drama. Because their scripts are at best dimly perceived, much research is designed to uncover the role of a species, or group of species, in a particular habitat, community or ecosystem.

What role does the spider play on its stage? What is its part in the network of interactions comprising the ecological web in which it lives? Certain basic roles the spider assumes frequently, so that in the repertoire of ecological plays the spider can be typecast to play one of a few fundamental characters. We also assume that spiders adhere, though with some improvisation, to a few simple scripts. Without these assumptions successful generalizing would be close to impossible. Our goal is to uncover the *hidden* scripts in order to understand the *interactive* dynamics with other players in the drama. But first we must complete a brief sketch of our leading actor.

The spider persona: a series of character sketches

A general portrait

Spiders are ubiquitous predators in terrestrial ecosystems. Spiders are generalist feeders that primarily attack insects, but also eat other arthropods, including spiders. They are even more strictly carnivorous than many other taxa of primarily predacious invertebrates such as centipedes and carabid beetles. Potent neurotoxins enable spiders to kill prey rapidly. Victims usually are smaller than or similar in size to the spider, but many spiders subdue prey several times their own mass.

Exceptions naturally exist. Some spiders break the rule of preying upon insects and other invertebrates by capturing an occasional vertebrate. The fishing spider *Dolomedes* even walks on water and dives to capture fish. A few spiders scavenge dead insects encountered in the search for live prey, and spiderlings of one species apparently derive nourishment from pollen grains ingested with the old web (Smith & Mommsen 1984). None, however, could reasonably be classified as primarily detritivore or omnivore. Occasionally spiders evolve into specialists. Bolas spiders, which eat only male moths, are extremists (Stowe 1986). These arachnids synthesize the sex pheromones of certain moth species and then capture approaching males by swinging a sticky droplet at the end of a silken thread. Little Miss Muffet thought she had problems. These New World spiders constitute the best example to date of convergent evolution between an arthropod and *Homo sapiens*: their common name reflects convergence in foraging tactics with cowboys of the South American pampas, and their accurate aim with a deadly sphere prompted Eberhard (1980) to name a newly described species (*Mastophora dizzydeani*) after a North American folk hero.

The list of exceptions could be continued. The uloborids have lost their poison glands. Some spiders have even abandoned the role of terrestrial carnivore. Two species, going even further than the fishing spider, have left the terrestrial habitat altogether. *Desis marinus* (Family Desidae) lives in the intertidal zone under kelp holdfasts (McQueen & McClay 1983, McClay & Hayward 1987), and *Argyroneta aquatica* (Family Argyronetidae) completes its life cycle submerged in freshwater habitats (Bristowe 1971). Aquatic oddities such as these have not led limnologists to categorize spiders as major components of the nekton, although the fishing spider *Dolomedes triton* qualifies as an important predator of the neuston, the community living on the surface film (Zimmermann & Spence 1989). Listing these exceptions to our initial character sketch, however necessary for accuracy, must not obscure our portrait of the typical spider. On balance, the spider merits the persona of generalist terrestrial predator in spite of the fascinating evolutionary byways a few taxa have pursued.

All spiders spin silk, though not all spin webs. Arachnida, the class to which spiders belong, takes its name from the hapless Arachne, whom Athena turned into a spider after the audacious girl presumed to be a better weaver than the goddess. The English word 'spider' is derived from the Old English verb 'spinnan,' to spin. 'Spinne,' the modern German word for spider, shares this root. Despite their ability to weave intricately beautiful webs, spiders are usually viewed in Western cultures as vermin, evil creatures to be avoided or preferably stepped upon.

Begrudging recognition that spiders kill equally abhorred insects earns them some toleration, but not much respect. E. O. Wilson (1978) raises fear of spiders to the level of an adaptive trait, suggesting that we are primed to acquire a strong, irrational fear of spiders early in childhood because of selective pressures exerted by spiders upon our ancestors. Such speculation appears silly when it is realized that not all cultures are as irrationally araneophobic as those of the West. Why is it that Western cultures abhor spiders; why is weaving almost universally considered women's work; why in English does the appellation 'spinster' have a negative connotation? These are questions best explored elsewhere (Weigle 1982).

Not all peoples treat spiders so poorly. Ananse, the Spider, is the hero of Ghanaian folk tales, having earned that role by outwitting God (Addo 1968). Many Native American cultures revere the spider's mastery of weaving to the point of elevating the spider to a basic creative force of nature. According to legends of the Pueblo people, '. . . in the beginning there was nothing but Spider Woman . . . no other living creature, no bird or animal or fish yet lived. In the dark purple light that glowed at the Dawn of Being, Spider Woman spun a line from East to West . . .' (Stone 1979). Spider Woman then created people from the clay of the earth, and to each she attached a thread of web connected to her.

Precedent suggests that the leading character in the ecological play I wish to explore is more likely to be an actress than an actor. The poetic tradition, common even among biologists, of referring to web-spinning spiders as feminine can be misleading. Both sexes spin webs, although mature males eventually cease web building and leave their web site in search of females. Male spiders usually mature before females and die earlier than their mates, who may live long enough to protect the egg sac and, in some species, care for the spiderlings. In many large orb-weaving species males are markedly smaller than females. All of these differences cause mature female web spinners to be more conspicuous and abundant.

The spider persona varies in relation to how silk is used to capture prey. Differences and similarities between spider families in the use of silk generally reflect patterns in habitat utilization and foraging behavior, which influence the spider's role in its ecological web. The clearest distinction is between the web spinners and the wandering spiders, which make silk but do not spin webs. Below I have sketched some portraits of the major families of the Order Araneae. Readers interested in exploring the biology of spiders in depth should consult the particularly comprehensive introductory overview of spider anatomy, physio-

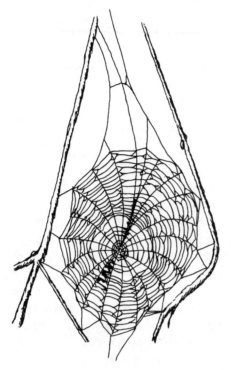

Fig. 1.1 Typical orb web of the family Araneidae.

logy, behavior and ecology by Foelix (1982). Several earlier works are also valuable (e.g. Nielsen 1932, Bristowe 1941, 1971, Witt, Reed & Peakall 1968, Gertsch 1979), and recently several excellent volumes on a variety of specialized topics have appeared (Witt & Rovner 1982, Barth 1985, Shear 1986, Nentwig 1987).

The web spinners

Web spiders weave a wondrous diversity of snares, from the most intricate of orbs to a trap recklessly reduced to a single silken strand. Ignoring the extreme specialists, we can recognize four basic web types: the orb, the tangle, the sheet and the funnel web.

Spiders in three of the major families weave orb webs. The largest family is the **Araneidae**, which first spin a supporting framework and then add a central catching spiral – the orb – made of silk containing minute sticky droplets (Fig. 1.1). The orb is delicate. Its structure is easily destroyed by wind, rain and struggling prey; and the threads soon lose

Fig. 1.2 A member of the Tetragnathidae (*Tetragnatha extensa*) with its orb web.

their stickiness. The spider renews the adhesive catching spiral and supporting radii (some species on a daily basis) by first ingesting the old web. Radiotracer studies show that components of the old web are quickly and efficiently recycled. Some orb-web spiders wait for prey at the orb's hub; others hide in a retreat connected to the hub by a signal line. The **Tetragnathidae** also weave orb webs, usually near or over open water (Fig. 1.2). Members of a third family, the **Uloboridae**, also

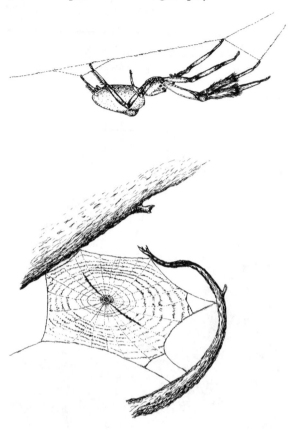

Fig. 1.3 A uloborid (*Uloborus glomosus*) with its orb of cribellate silk. The spider hangs underneath the horizontally hung web.

employ an orb web to capture prey. Uloborids do not make sticky silk, but instead cover the threads of their catching spirals with a fine meshwork of specialized cribellate silk (Fig. 1.3). They belong to the cribellates, the 'hackle-band' weavers, a group that includes several other relatively small families.

Tangle webs are irregular mazes woven by the **Theridiidae** (Fig. 1.4). The network may be placed high in shrubs, trees or under rock outcrops. Other species locate the snare close to the ground, with elastic, sticky threads extending from the maze to the ground where they capture crawling prey. Theridiid species wait in their snare for prey, sometimes in a retreat they have constructed, or under an adjoining leaf or in an adjacent crevice.

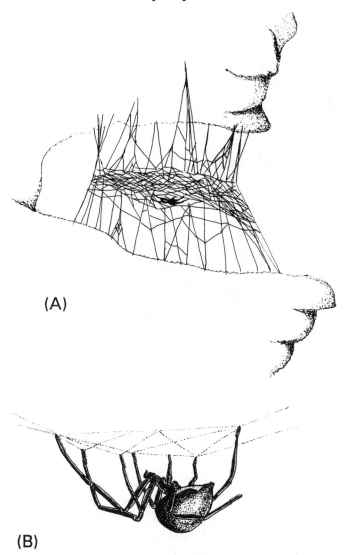

(A)

(B)

Fig. 1.4 Two tangle-web weavers of the family Theridiidae: (A) *Steatoda* sp., (B) A female black widow, *Latrodectus mactans*.

Spiders in the Family **Linyphiidae** build sheet webs of entirely non-sticky silk. The basic linyphiid web is a horizontal sheet with scaffolding above and below (Fig. 1.5). Each species modifies the sheet, from a bowl or hammock through relatively flat sheets to finely constructed domes. Linyphiids hang underneath the sheet waiting for prey, which they pull

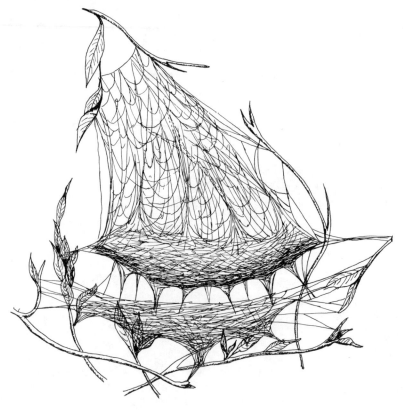

Fig. 1.5   Web of the linyphiid *Frontinella pyramitela*, the bowl and doily spider. The spider (not depicted) hangs underneath the 'bowl' and is protected from below by the 'doily.'

through the sheet after the insect has become momentarily entangled in the upper network of threads. The web is repaired as tears develop, but is not renewed daily.

The funnel-web spiders (Family **Agelenidae**) also make a sheet web, but their utilization of the sheet differs markedly from the linyphiids. Agelenids are closely related to the wolf spiders (Lycosidae) and resemble them in both morphology and movement. The typical agelenid constructs a funnel at the edge of the web inside a curled leaf or a crevice (Fig. 1.6). When prey lands on the sheet the agelenid runs rapidly across the top of the sheet, captures the prey and returns with it to the retreat.

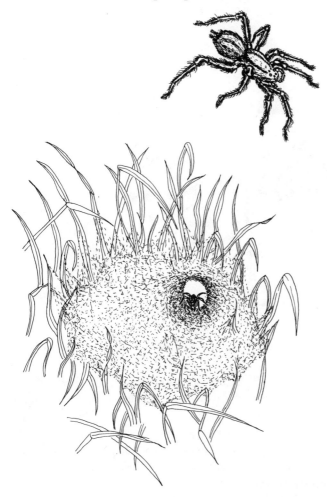

Fig. 1.6 Agelenopsis naevia, a funnel-web spider of the family Agelenidae.

Wandering spiders

Not all spiders use a web to capture prey. The wandering spiders restrict their use of silk to the dragline they often trail behind them as they move about, to protecting the eggs or in some cases to lining their retreat. These spiders run after their prey or wait for it in ambush.

Crab spiders (Family **Thomisidae**) wait for prey in flowers, on leaves or on tree trunks (Fig. 1.7). Because they rely on ambush, thomisids forage in a manner similar to the web spinners, which frequently are

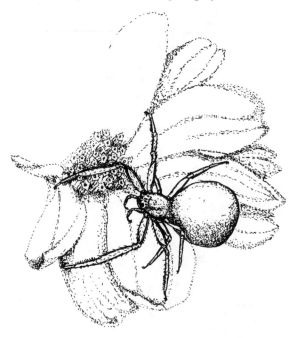

Fig. 1.7 A gravid crab spider, *Misumena vatia* (Thomisidae), waiting in ambush on a flower.

categorized as 'sit–and–wait' predators. Crab spiders derive their name from the ability to scurry sideways or backwards. Thomisids also vaguely resemble the crustacean in appearance, with a flattened body and powerful front legs that have become rotated so that the lateral surface is almost dorsal. These spiders move sideways and backwards with the same agility as crabs.

Jumping spiders, members of the **Salticidae**, are the most active of the wandering spiders. Salticids stalk their prey and pounce on them, using saltatory skills directed by the best eyes in the spider world (Fig. 1.8). In contrast to the eyes of web builders, the primary eyes of jumping spiders can form sharp images and the binocular vision of the secondary eyes enables salticids to estimate distances. Salticids hunt only during daylight and apparently are unable to capture prey in the dark.

Wolf spiders (**Lycosidae**; Fig. 1.9) do not roam in packs nor do they run down their prey as frequently as their name implies (Stratton 1984). Although very active on the soil surface and in the leaf litter of forests and meadows, lycosids often wait quietly in ambush for their prey. Their

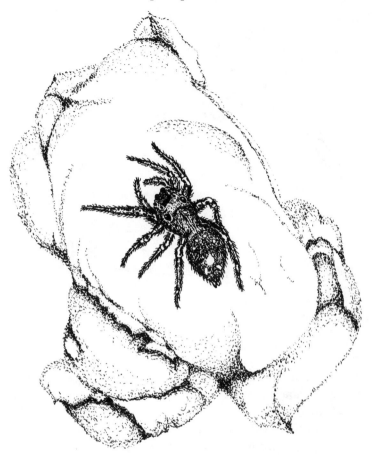

Fig. 1.8 The jumping spider *Phidippus audax* (Salticidae).

eyes cannot form images as clear as those of salticids, but their eyesight is better developed than that of the web builders. Several other families of spiders also do not spin webs. The **Clubionidae** and the **Oxyopidae** are two additional types of wandering spider that we will meet later (Fig. 1.10). Clubionids are short-sighted rapid runners that usually spend the day in a silken retreat and hunt at night. Some wander on the soil surface and others range over vegetation. The oxyopids, or lynx spiders, have become specialized to living on vegetation. Their well-developed eyesight enables them to move quickly over leaves and stems, though some oxyopid species ambush their prey.

Fig. 1.9 A typical wolf spider, *Lycosa communis* (Lycosidae).

More primitive spiders

The families just described belong to the Suborder Araneomorphae, which includes most species in the Order Araneae. The more primitive Suborder Mygalomorphae includes the spectacularly large spiders of the Family **Theraphosidae**, which Americans have named 'tarantulas' (the European 'tarantula' is a lycosid); the trap-door spiders of the Family **Ctenizidae**, and a few smaller families. Mygalomorphs are generally larger than the more recently evolved Araneomorphae and include the largest and longest-lived spiders. Most mygalomorphs are ground dwellers. They hunt from subterranean burrows or utilize sheet or tubular-shaped webs placed on or near the ground.

Major themes

Ecological webs

The ecological web is a widely used metaphor that is often imprecisely defined. Recently Andrewartha & Birch (1984) proposed to limit its meaning by defining it as the indirectly acting components of the environment of a particular species. In contrast to their restrictive view, I perceive the ecological web as the network of both direct and indirect interactions that link together the organisms in a community, a tangled maze of relationships resembling Darwin's (1859) image of 'plants and

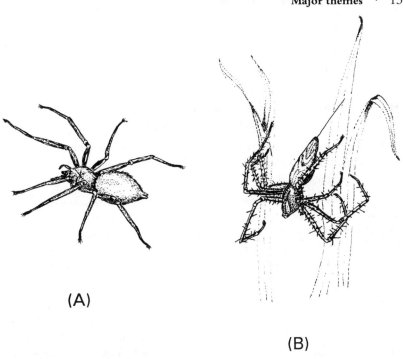

(A)

(B)

Fig. 1.10   Two other important families of wandering spiders: (A) Clubionidae
(*Clubiona obesa*), (B) Oxyopidae (green lynx spider, *Peucetia viridans*).

animals remote in the scale of nature, . . . bound together by a web of
complex relations.'

I have three goals in developing the role of spiders in ecological webs.
First, I wish to characterize the basic properties of a group of organisms
that together constitute one model of the generalist terrestrial predator.
Secondly, and as an inseparable part of the first objective, I wish to
examine the web of interactions between spiders and other organisms in
terrestrial communities. These generalizations will increase our understanding of a major type of organism, the spider, and will also be directly
relevant to several general issues in ecology, such as the importance of
competition among resource-limited organisms. Thirdly, I wish to
explore in depth the design and interpretation of field experiments that
ecologists have performed to test hypotheses about spiders. This third
goal reflects my belief in the potential of field experimentation to address
issues at several levels of ecological organization, and is conveniently
explored with spiders because of their suitability to the experimental

approach. As ecologists increasingly rely on field manipulations, it becomes important to examine periodically both the shortcomings and successes of field experiments in order to strengthen future research.

Model organisms and model communities

Spiders are abundant and ubiquitous predators in terrestrial ecosystems. Any understanding of the way in which carnivores function in terrestrial systems must incorporate spiders, which led me to propose the spider as a model terrestrial predator (Wise 1984a). Clearly, however, spiders are one of several possible models; among invertebrates, centipedes or predacious beetles are equally good candidates. Many vertebrate models would have their advocates, and those who study parasitoids could reasonably argue that generalist feeders certainly do not qualify as *the* model predator for terrestrial ecosystems. No single taxon can be anointed the sole model terrestrial predator. In fact, some spider families differ so much in how they forage and utilize their surroundings that it may prove risky to generalize about the role of *the spider* in terrestrial communities. Competitive interactions, impacts upon prey populations and susceptibility to natural enemies may differ so much between web builders and wandering spiders that the roles of these two groups in ecological webs would have to be assessed separately.

Such fascinating but vexing variation on major themes highlights the need to identify model systems in ecology, an initial step that more reductionist sciences take almost without conscious effort. But how should we define our model systems? This is a central question confronting those who advocate a more mechanistic approach to the study of ecological communities (McIntosh 1987). Focus on competition has fostered a tendency to define communities horizontally, often along taxonomic lines. Examination of the role of competition in spider communities, and extrapolating from the behavior of this model community to infer the importance of competition among communities of terrestrial carnivores in general, would be in this mainstream tradition. The more daring would venture even farther. By choosing a taxonomic orientation, I have committed myself to this course, at least to the point of examining competition in spider communities. An additional impetus to examining a community that has been defined taxonomically is the conviction that we can successfully generalize about organisms as similar to each other as spiders; if it turns out that we cannot, we may be in more trouble than even the most pessimistic among us lament.

The spider community *per se* is of limited interest, though useful preliminary insights can be gained by focusing on interactions within a limited grouping of species. The ultimate goal is to understand the workings of larger assemblages of interacting organisms. Even to understand spiders one needs to look at the other organisms with which they interact. This is not a new nor profound insight. Nevertheless, a strong tradition exists in which different communities (i.e. spider communities, bird communities, plant communities, etc.) are compared in order to uncover meaningful patterns about how ecological systems function. Such comparisons, though sometimes enlightening, yield only glimpses of the processes that define the fuller system to which the taxonomically delineated community belongs. Ecologists have realized this, but continue to make the comparisons, primarily because more complex systems are discouragingly difficult to study. Redefining taxonomically restricted communities as guilds or assemblages does not remove the problem, but does serve to emphasize that a taxonomically defined grouping is not what ecologists really mean by *community*.

Spiders comprise a node in a complex tangle of interactions – the real communities of ecology. In this view spiders themselves can no longer constitute a model community because they merely are part of a more complex system. Intensive study of a few complex systems hopefully will reveal generalizations applicable to similar systems. The detailed case studies then become the appropriate models for the general class of similar communities. One fruitful approach to unravelling the complexity of these communities is to experimentally perturb major components. Manipulations are most rewarding when the properties of the major system components are understood. Generalizations derived from studies of limited numbers of spiders enable us to characterize spiders as a group, a mandatory first step to generalizing about the role of spiders in the complex webs of terrestrial communities.

Field experiments in community ecology

Field experimentation is now generally accepted as a powerful tool for testing ecological hypotheses (e.g. experiments reviewed by Connell 1983, Schoener 1983, Sih *et al.* 1985, Hairston 1989). Field experimentation must become even more widely employed in the study of ecological webs for several reasons: (1) Field experiments are the most direct way to test hypotheses about patterns derived solely from observational, correlative data. Such patterns are themselves often termed 'natural experiments,' which implies that they can legitimately substitute for controlled field manipulations (e.g. Cody 1974, Diamond 1986). The two approaches are not equivalent. Natural experiments can suggest hypotheses but cannot be considered direct tests of hypotheses (Wise 1984a, Hairston 1989). (2) Long-term perturbation experiments offer a powerful approach for uncovering indirect interactions in communities (Bender, Case & Gilpin 1984). (3) In addition to testing specific hypotheses, field experimentation is often the most direct way to uncover unexpected relationships in complex ecological systems and can serve as a valuable hypothesis-generating device (Chew 1974, Usher, Booth & Sparkes 1982, Brown *et al.* 1986, Diamond 1986).

Untangling the web

How to begin untangling strands of the spider's ecological web is a vexing question. The goal of this book is to synthesize and place in perspective the results of numerous research programs on spiders, which often differ in underlying rationale, approach and scope. Any research program of necessity reflects a particular ecological world view and thus addresses questions framed in the context of a particular notion of how ecological communities are structured. A clear example is the focus of many ecologists upon questions of resource limitation and competition. Thus the available data reflect the historical development of ecology. Furthermore, the metaphor of the role of a model organism in an ecological web by itself dictates the approach to be taken in synthesizing available information.

The metaphor of spiders occupying an ecological web conjures up an image of an abstract arachnid connected by silken causal strands to a multitude of creatures, with the number and kinds of silk varying in reflection of the intensity and directness of the interactions. At the end of our journey this metaphor may be reduced to a jumbled sticky tangle of broken threads, but at least the image provides an initial framework for beginning the task at hand. First we will dissect the model creature itself by determining the extent to which spiders interact with each other. One class of central issues is whether intra- and interspecific competition between spiders constitute major connections in their ecological web. Related threads are the role of predation between spiders, and the impact of competition for resources between spiders and other carnivores. Following vertical strands through our metaphor leads to questions of limitation of spiders by natural enemies and the impact spiders exert upon their insect prey. Along the way we will discover strands that stretch and spiral back; tugging on a nearby thread may spread vibrations throughout the web, leading to changes at nodes far removed. Thus the metaphor will lead to a concluding examination of indirect effects in the ecological webs of spiders.

Synopsis

If the terrestrial world is a stage, then any predator as abundant and ubiquitous as the spider must be a major character in the ensuing ecological and evolutionary dramas. Following a description of the generalized spider persona, portraits of some of the major spider families are sketched. Differences in how spiders use silk to capture prey are often correlated with differences in foraging behavior and habitat utilization. The basic division is between web builders and wandering spiders.

There are four basic web types: orbs, tangles, sheets and funnel webs. Three major families construct orb webs (Araneidae, Tetragnathidae and Uloboridae). The largest family is the Araneidae, who rely on a central spiral made of viscid, sticky silk to capture prey. Tetragnathid webs are similar, but orbs of uloborids are made of specialized, non-sticky cribellate silk. Theridiids spin irregular tangles that may incorporate some strands of sticky silk. Linyphiids hang underneath a sheet made of non-viscid threads. The Agelenidae, or funnel-web spiders, also construct a sheet, but they run rapidly on top of it to capture prey, which are then brought back to the funnel retreat to be eaten.

Wandering spiders have abandoned the use of a snare to capture prey. Crab spiders (Thomisidae) wait patiently in ambush, either within a flower, on a leaf or on the bark of a tree. Jumping spiders (Salticidae) stalk prey and pounce on their victims. Lycosids, the wolf spiders, are active wanderers but do not have as keen eyesight as the salticids; they are more likely to wait in ambush than to run after their prey. Clubionids have poor eyesight and hunt at night, whereas the lynx spiders (Oxyopidae) are specialized for active daytime hunting on vegetation.

The following chapters develop themes that reflect the three primary goals behind the writing of this book. The first is to characterize the spider as a model terrestrial predator. The second goal is to clarify what we know about interactions between spiders and other organisms in terrestrial communities. Stated more metaphorically, what are the roles of spiders in their ecological webs? Related to these topics are the more general issues of what constitutes a model organism or a model community, and how we should investigate complex ecological systems. The third goal is to examine critically field experiments involving spiders. This objective springs from my conviction that field experimentation is integral to research at both the population and community level. Fortunately, many families of spiders have proven amenable to the experimental approach. I will weave all of these themes together in my attempt to untangle the spider's ecological web.

2 · *Hungry spiders*

Food limitation of terrestrial carnivores

The concept of resource limitation is so central to ecological thinking that it might seem unnecessary to justify examining the impact of prey supply upon spider populations. However, attempts to uncover broad patterns in resource limitation frequently have generated controversy. One that directly engulfs spiders started with a brief, carefully argued communication by Hairston, Smith & Slobodkin (1960). They concluded that the carnivore trophic level of terrestrial ecosystems is 'resource-limited in the classical density-dependent fashion.' In particular, they argued that predators are food limited, and that competition occurs on this trophic level. Hairston *et al.* did not propose that every group of terrestrial carnivores is food limited. Nevertheless, because spiders are major terrestrial predators, it follows that a shortage of prey should frequently affect spider densities. As a model terrestrial predator the spider cannot escape the controversy created by the sweeping predictions made by Hairston *et al.*; indeed, continuing interest in testing their predictions in conjunction with disagreement over the prevalence of competition has already drawn spiders into the fray (Wise 1975, Schoener 1983a, 1986a).

Prey is conventionally defined to be a limited resource if an increase in the prey supply increases the predator survival and/or fecundity. If increases in one or more of these parameters cause the average population density of the next generation to increase, the population is food limited. Thus, food limitation is defined ultimately in terms of population density, but evidence that food is limiting frequently comes from within-generation measurements of individual survival rates or fecundity. Effects on growth and foraging behavior also are taken as indirect evidence because they often affect survival or reproduction. Because of the potentially complex interactions between factors limiting population density, effects of prey shortages on parameters such as fecundity cannot be considered conclusive evidence of food limitation at the population level. Nevertheless, other limiting factors will unlikely compensate perfectly for increases in survival or fecundity that a prey increase may produce in a spider population; thus, observed effects of food supply on basic demographic parameters can reasonably be taken as good indirect evidence of food limitation at the population level. Firm proof, though, only comes from demonstrating that a sustained increase in prey supply causes a higher density of the predator.

Prey limitation of carnivores is usually defined, often implicitly, in terms of joules of prey available, perhaps because carnivores are less likely than autotrophs and herbivores to be limited by nutrient shortages (shades of Hairston *et al.* 1960), and also because joules are the currency of most foraging theory. However, food limitation in spiders may involve more than a scarcity of joules. Greenstone (1979) argues that the lycosid *Pardosa ramulosa* selects prey in a manner that optimizes the proportion of essential amino acids in its diet. This result is surprising. One would predict that the exact composition of the diet should not make a difference for polyphagous predators. Rearing studies with other *Pardosa* species demonstrate that a mixed diet in the laboratory is necessary for normal growth and reproduction (Miyashita 1968, Van Dyke & Lowrie 1975). Greenstone argues from this result that natural selection could have favored selective polyphagy in *Pardosa*. Accumulating evidence shows that other types of spiders are also healthier in the laboratory when fed more than one species of prey (i.e. Lowrie 1987, Riechert & Harp 1987), though it does not follow that these species in nature select numbers of prey on the basis of their amino acid content. *Pardosa ramulosa* fed primarily upon three aquatic insect species present at the margins of pools in the salt marsh investigated by Greenstone. The spectrum of available prey in this habitat is more restricted than for most habitats inhabited by wolf spiders. Feeding behavior that optimizes nutrient content of the diet may evolve in environments with such low prey diversity, but not in most situations inhabited by spiders. Possible nutrient limitation and selective polyphagy based upon nutrient content of prey require additional research before the issue will be resolved for spiders.

The diets of spiders clearly do not precisely mimic the proportions of insects in the habitat (e.g. Kajak 1965, Uetz & Biere 1980, Castillo & Eberhard 1983, Nentwig 1980, 1985b); nevertheless, proportions of different types of prey in the diet, and particularly changes in diet, often largely reflect prey availability (Turnbull 1960, Riechert & Łuczak 1982, Riechert & Harp 1987). Although differences might reflect varying nutrient contents of prey, such differences can be explained most readily by (1) rejection of certain prey because they are toxic, dangerous or difficult to handle (e.g. Bristowe 1941, Nentwig 1983); (2) the initial avoidance of novel prey (Turnbull 1960b, Riechert & Łuczak 1982); (3) the degree of satiation of the spider (Nakamura 1987, Riechert & Harp 1987); or (4) different sampling methods used by ecologists and spiders.

Practically all evidence of food limitation in spiders treats prey as resource packets measured as numbers, milligrams or joules of prey. Conclusions concerning the extent of food limitation are based upon evidence ranging from field correlations to controlled field experiments. The apparent importance of food limitation in the evolution of spiders constitutes the broadest, but most indirect, evidence.

Evolutionary arguments

Metabolic rates

Physiological responses of spiders to starvation suggest they have experienced food shortages frequently throughout their evolutionary history. Older instars of numerous species can survive long periods of starvation. For example, Anderson (1974) found that under laboratory conditions adults of the lycosid *Lycosa lenta* could survive starvation an average of 208 days, and adults of a web-building species an average of 276 days. Spiders survive starvation by maintaining a relatively motion-less sit-and-wait foraging strategy, and by decreasing their basal metabolic rate (Itô 1964, Miyashita 1969, Nakamura 1972, Anderson 1974, Tanaka & Itô 1982, Tanaka, Itô & Saito 1985).

Several investigators have reported that resting metabolic rates of spiders are lower than those of comparably sized invertebrates, which suggests that spiders may experience particularly severe prey shortages (Anderson 1970, Greenstone & Bennett 1980, Anderson & Prestwich 1982). However, Markezich (1987) doubts that spiders have unusually low rates of metabolism. Because the standard regression relating size and metabolic rate of invertebrates to which spiders have been compared is derived from non-starved poikilotherms, Markezich attributes the lower weight-specific rates observed by Anderson and others to the fact that spiders had been starved for many days before oxygen consumption was measured. In support of his contention that metabolic rates of spiders are higher than widely believed, Markezich cites his studies, and those of others, in which field-collected, non-starved spiders do not exhibit respiratory rates that are lower than predicted for poikilotherms of their size. It appears that spiders may not be unusual in their physiological adaptations to starvation. In order to resolve this issue, we need studies in which the responses of spiders to starvation are compared to those of other arthropod predators of similar size. The clear generalization remains, though, that spiders markedly reduce their resting metabolic rate when deprived of prey.

Foraging constraints: temporal and spatial patterns

In addition to physiological adaptations to food shortages, temporal and spatial patterns in spider foraging suggest that prey scarcity has plagued spiders for millennia.

Olive (1981) argues that phenologies of orb-weaving spiders evolved in response to foraging constraints imposed by a limited food supply. He constructed a model in which energy return is a function of spider size, habitat type and time of the season. He then compared the seasonal size distribution of mature spiders for several species with the pattern predicted by the model, which incorporated the assumption that the phenologies evolved to maximize the average rate of energy gain throughout the life cycle. For all but one type of site, the season of maximal energy return for large species is late August, which is when they mature. The minimum body size at which peak energy return shifts from autumn to spring agrees well with the size distribution of autumn-maturing orb weavers. Olive argues that exceptions to the predictions can be explained in terms of the natural history of the most abundant prey in these sites.

Differences between the sexes in surviving harsh physical conditions provide indirect evidence of food limitation. In a five-year study of the sheet-web spider *Pityohyphantes phrygianus*, Gunnarsson (1987) discovered that more males than females died during three of the winters. Through field and laboratory experiments he confirmed that males were more susceptible to cold temperature. Possibly males die at a higher rate because they are smaller than females; Gunnarsson, however, hypothesized that pressures of sexual selection actually favor large size in males, leading to a diversion of energy from the synthesis of glycerol and other molecules that depress the freezing point of the hemolymph.

Riechert & Łuczak (1982) argue that the evolved 'sit-and-wait' strategy employed by many spiders is evidence of the evolutionary impact of food limitation. Foragers confronted with a prey shortage have two options: remain and wait for an increase, or search for a more productive microhabitat. Norberg (1977) developed a general model which predicts that a decrease in prey abundance can favor the sit-and-wait strategy. Janetos (1982a,b) attempted to explain differences between araneids and linyphiids in residence times at web sites in terms of differences between the two families in the energetic costs of web construction and variance in prey availability. His model, though, does not explicitly incorporate the overall rate of prey supply in the habitat.

Caraco & Gillespie (1986) proposed a risk-sensitive foraging model that incorporates both variance in prey capture rates and average prey availability. Their model predicts that when temporal variability in prey supply is sufficiently high, overall scarcity of prey in the habitat can favor the sit-and-wait foraging mode. Differences in foraging behavior between two populations of long-jawed orb weavers (Tetragnathidae) that inhabit areas with markedly different average prey abundances support this prediction (Gillespie & Caraco 1987).

Long before the birth of optimal foraging theory, Kirby & Spence (1815; cited by Cherrett 1964) stated that spiders place their webs where prey are particularly abundant. Many web builders apparently follow this conventional wisdom. The desert spider *Agelenopsis aperta* selects sites for its funnel web where prey is most abundant and where thermal conditions will allow the most time for foraging (Riechert & Tracy 1975; Chap. 7). Dabrowska-Prot, Łuczak & Wojcik (1973) found a correlation between abundance of web builders and prey density in an ecotone between meadow and alder forest. Laboratory studies and field experiments conducted in enclosures have shown that web-building spiders leave areas of low prey abundance and tend to remain where rates of prey capture are greater (Turnbull 1964, Gillespie 1981, Olive 1982, Vollrath 1985).

Janetos (1986) and Riechert & Gillespie (1986) review the accumulating studies of the factors that affect both selection and abandonment of an occupied site. In addition to prey availability, vegetation structure and exposure to insolation and wind all appear to play a role. Janetos (1986) argues that in order to understand web-site selection, we should be addressing the question of what factors influence desertion of a web site. Evidence suggests that rate of prey capture often is important, but counteracting factors, such as increased exposure to predation during dispersal to a new web site, may favor tenacity to a site that provides fewer prey than the spider is capable of capturing and assimilating.

Many wandering spiders also select microhabitat on the basis of prey abundance. The crab spider *Misumena vatia* tends to occupy milkweed umbels that have the most insect prey (Morse & Fritz 1982), which probably results from spiders leaving inflorescences on aging stems and moving to younger stems, whose flowers attract more prey. The matching of spider and prey distribution is by no means perfect. Within a period of a day, approximately a third of the thomisids did not leave senescent stems, and a quarter of the spiders that had been transferred experimentally to high-quality stems moved. There was no evidence that these crab spiders moved regularly between milkweed clones several meters apart. Some of this variability, and also the tendency of *M. vatia* to move only small distances, can be accounted for by unpredictability of visits to flowers by insects (Morse & Fritz 1982).

Upon maturing, females of the desert wolf spider *Lycosa santrita* move from grass to bare areas that have a higher prey productivity (Kronk & Riechert 1979). A field experiment, in which individually caged spiders were given known amounts of prey, confirmed that *L. santrita* can utilize the greater productivity of the open habitat during the egg-ripening period. Edgar (1971) found that *L. lugubris* with eggs moves from shaded areas into forest clearings where prey were more abundant.

Edgar (1971) and Kronk & Riechert (1979) caution that food availability may not be the only factor causing mature female lycosids to shift microhabitat. Both *Lycosa* spp. moved into sunnier microhabitats. Because lycosid females carry their egg sacs attached to their spinnerets, the higher temperatures of more exposed areas could increase female fitness by accelerating egg development. Thus the observed shift in habitat utilization of these two *Lycosa* spp. might not be solely an adaptation to improve foraging efficiency. Nørgaard (1951), for example, demonstrated that a bog-inhabiting wolf spider moves upwards in the *Sphagnum* mat in order to expose her egg sac to higher temperatures.

Patterns of microhabitat utilization by young lycosids also have more than one interpretation. Unlike their mothers, immature desert wolf spiders do not inhabit the more productive open areas. Although Kronk & Riechert (1979) suggest that the prey-capture efficiency of smaller lycosid instars may be higher in the grass, they also speculate that this microhabitat may provide protection from pompilid wasps. In contrast to the desert lycosid, young *L. lugubris* remain in the clearings and their mothers return to the more shaded microhabitat. Edgar (1971) suggests that female emigration is adaptive because it reduces cannibalism by females upon the young. This hypothesis requires group selection because females potentially can kill the young of unrelated spiders as well as their own progeny. A more parsimonious hypothesis is that other selective constraints have caused females to return to the forest, and avoidance of predation by mature females plus the warmer temperatures and greater prey productivity of the clearings favor spiderlings that remain in these more open areas.

Food limitation over ecological time

Inferring the current importance of a process such as food limitation from evidence of its evolutionary impact is risky for at least two reasons. First, different selective factors can favor the same character. For example, although foragers may sit and wait for prey because prey are scarce, factors other than prey abundance also influence foraging mode (Schoener 1987). Secondly, adaptation to a limiting factor over evolutionary time may permit a species to escape from that factor over ecological time. One could argue that past prey shortages have produced adaptations in spiders that remove food supply as a limiting factor for current spider populations. However, evolved behavior that maximizes prey intake in an environment in which prey availability varies spatially and temporally does not guarantee that the forager will escape food limitation. Spiders may still not capture enough prey to achieve maximum possible rates of growth and reproduction. Several lines of evidence indicate that despite adaptations to survive food shortages and locate in areas of higher prey abundance, food supply continues to be limiting for spider populations over ecological time scales.

Indirect evidence

Variation in natural populations
Extensive variation in size at maturity within populations suggests that many spiders do not capture enough prey to achieve maximum rates of growth (e.g. Kajak 1967, Wise 1983). During a two–year study of the crab spider *Misumena vatia*, 20% of the population did not capture enough prey to reproduce (Fritz & Morse 1985). Similarly, Horton & Wise (1983) discovered that during a particularly dry year, most females of the orb weaver *Argiope aurantia* did not attain reproductive size by the end of the season, presumably because of a shortage of prey resulting from a drought.

More direct evidence of food limitation comes from correlations between prey abundance and parameters such as growth, fecundity and population density. Hubbell (1932) discovered that an unusually dense

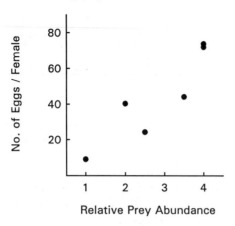

Fig. 2.1 Correlation between fecundity of *Erigone arctica* and abundance of a major prey species, the collembolan *Hypogastrura viatica*. Relative prey abundance is expressed as a ranking: $1 = 1–160/m^2$; $2 = 161–400/m^2$; $3 = 401–1600/m^2$; $4 = 1601–16000/m^2$. (Based upon data in Table II from van Wingerden 1978.)

aggregation of orb weavers (Tetragnathidae and Araneidae) along the bank of a Florida lake was associated with a massive autumn emergence of a midge. Miyashita (1986) followed growth and reproduction of the orb weaver *Nephila clavata* in two 12 × 12 m plots in Japan for two years. In the first year of the study a higher prey abundance in one plot was correlated with a faster rate of development, larger size and higher fecundity of *N. clavata*. These parameters did not differ between plots in the second year, when prey abundances also did not differ significantly. Such a small sample size provides only suggestive evidence. Stronger evidence comes from a study of a small linyphiid, *Erigone arctica*. Fecundity of this sheet-web spider, which is common in periodically flooded dune habitats in the Netherlands, appears to vary in response to the availability of a major prey species, the collembolan *Hypogastrura viatica* (van Wingerden 1975, 1978). Van Wingerden (1978) guardedly concluded that over the 3.5 years of his study there was a 'weak correlation between the egg production per female and the density of *H. viatica*.' A stronger conclusion is justified. Calculation of the correlation coefficient from data presented in Table II of his 1978 paper reveals a convincing, statistically significant positive correlation between prey abundance and the fecundity of *E. arctica* ($r = 0.9$, $p = 0.01$; Fig. 2.1).

Such effects on growth and fecundity should translate into changes in population densities of spiders. Cherrett (1964) found that the density of

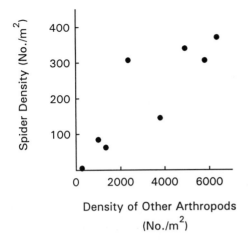

Fig. 2.2 Relationship between densities of spiders and densities of other arthropods across eight different habitats, ranging from bog and *Calluna* heath to grazed, enclosed and cultivated areas on limestone. (After Cherrett 1964.)

linyphiids over a range of moor habitats was positively correlated with the density of other arthropods ($r = 0.86$, $p < 0.001$, d.f. $= 6$; Fig. 2.2). Abundance of insect prey may determine the density of wolf spiders in a coastal landscape of the Baltic Sea. Schaefer (1972) uncovered a statistically significant correlation ($r = 0.86$, $p < 0.001$; d.f. $= 15$) between a measure of abundance of all wolf spiders, based upon pitfall trapping, and an index of prey abundance (Fig. 2.3). Nentwig (1982) reviewed several studies in which pitfall traps had been used to capture spiders and their potential prey. Collembola and Diptera captures were positively correlated with numbers of spiders trapped. Nentwig found similar positive correlations between numbers of spiders and potential prey in traps that he had placed in a moor and wet meadow ($r = 0.56$ and 0.83, respectively; $p < 0.001$).

The preceding indirect evidence for food limitation seems convincing, but purely correlative evidence is always open to alternative interpretations. For example, could not the clear correlation between prey abundance and fecundity of *E. arctica* have resulted from variation in a third factor, perhaps temperature, that positively affected fecundity and survival of both spider and its prey? Cherrett (1964) points out that the correlation he observed could reflect a positive impact of vegetation structure upon the density of both spiders and other arthropods, with no causal connection between numbers of spiders and their prey. This is

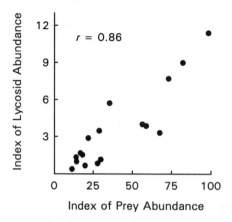

Fig. 2.3 Correlation between the availability of potential prey and the abundance of wolf spiders collected by pitfall traps in 17 different areas in salt meadows, heath and dry dunes along the coast of the Baltic Sea. Index of lycosid abundance equals the volume (cm³) collected per trap. The prey index is based upon standardized collections by sweep netting and pitfall trapping, expressed as a percentage of the maximum prey volume found in the sampled areas. (After Schaefer 1972.)

possible, since the correlated data points are spider and arthropod abundances in distinctly different habitats. Numbers of spiders and their prey sampled by pitfall traps are both highly correlated with temperature (Nentwig 1982), reflecting the fact that the traps are measuring both density and activity of the organisms. Thus the tight relationship between numbers of spiders and prey trapped may not have resulted from a direct causal connection between the dynamics of prey and spider populations. In the absence of experimentation such doubts will always remain.

Laboratory–field comparisons
One approach to controlling confounding variables has been to compare growth and fecundity of laboratory-reared spiders with field-collected specimens. Edgar (1969) concluded that the wolf spider *Lycosa lugubris* feeds only infrequently in nature, since his calculated field rate of one prey organism per day is less than the rate that can be achieved in the laboratory. He quotes Savory's (1964) view that '. . . many a spider is overfed if given a fly daily.' Population censuses of several families of wandering spiders have revealed low feeding frequencies (<10% observed feeding; Nyffeler & Breene 1990). Growth rates of spiders in

nature generally are lower than under conditions of superabundant prey in the laboratory. Hagstrum (1970) found that laboratory-reared wolf spiders *Tarentula kochi* were 27% heavier than field-collected specimens of the same carapace width. Kessler (1973) discovered that four *Pardosa* species (Lycosidae) fed a superabundance of fruit flies in the laboratory matured larger and had a higher fecundity than field-collected spiders. Anderson (1974) compared the weights of laboratory-reared and field-collected individuals of a wandering spider, *Lycosa lenta*, and a web spinner, *Filistata hibernalis*, and concluded that both experience severe food shortages in nature. Physical conditions in these laboratory studies, in addition to increased prey, may have favored growth, so that as evidence of food limitation in nature the results are highly suggestive but not conclusive. They are no substitute for results obtained from controlled field experiments.

Field experiments

Field experiments provide direct evidence that food supply is a limited resource for web-building spiders. In these studies investigators manipulated the rate of food supply under natural conditions and observed changes in web-site tenacity, growth or fecundity. Web-building spiders offer two distinct advantages to the experimental ecologist. First, the sedentary habits of web spiders eliminate the need to build restraining barriers. Caging web-building spiders is undesirable because cages (1) decrease normal prey availability, thereby imposing food limitation on a system that may not have been prey-limited; (2) decrease exposure to natural enemies, thereby elevating survivorship artificially; (3) prevent emigration, which is a likely response to prey shortages; (4) possibly elevate population density artificially, leading to artifacts such as increased agonistic interactions and cannibalism; and (5) may significantly alter the physical environment.

The second gift of web weavers to the experimentalist is the web itself, because one can increase the rate of prey supply to specific spiders simply by dropping prey into the web. I used this technique to demonstrate that prey supply is a limiting factor for the filmy dome spider *Neriene radiata* (= *Linyphia marginata*) (Wise 1975). Most populations of this linyphiid exhibit a variable life history pattern (Wise 1976, 1984b). A wide range of immature stages overwinter. The larger ones mature early in the spring, reproduce and die. Their progeny hatch within a few weeks, develop rapidly and become mature during August, but at a smaller size than spring-maturing adults. Offspring of these adults overwinter as the smaller instars and become reproductive late in the following spring. These spiders' progeny do not mature that season, apparently because a developmental switch retards developmental rate of spiderlings hatching from egg sacs laid late in the spring (Wise 1987). As a consequence they have a longer growth period and thus mature at a larger size than the rapidly developing, summer-maturing spiders. Because of the variation in developmental rates, late-summer populations of the filmy dome spider contain adults and a range of immatures; in contrast, early-spring populations contain only immature stages.

Filmy dome spiders appear to be more food-limited during summer than spring, based upon an experiment conducted with spiders inhabiting ground juniper bushes in a Michigan deciduous forest (Wise 1975; Chap. 5). Providing summer females with additional prey more than doubled their egg production, from 18 ± 10 to 46 ± 10 eggs per female. Supplementing the prey of juvenile spiders during the summer (offspring of spring females) increased their growth rate by 30% as measured by increase in weight. The relative shortage of prey also likely decreased the rate of molting (Martyniuk & Wise 1985, Wise 1987), but this was not measured. Lowered rates of growth and development of juvenile *N. radiata* resulting from prey shortages potentially affect population density by reducing future fecundity due to the strong correlation between size at maturity and egg production (Wise 1976), and by decreasing survival over the winter due to the higher overwinter-mortality rate of earlier instars (Martyniuk & Wise 1985). Prey appeared to be more abundant during the spring. Supplementing the prey of females increased fecundity by only 15%, a statistically insignificant increase ($p(t) > 0.15$; Chap. 5).

Filmy dome spiders hide their egg sacs in the leaf litter, a feature not so attractive to the experimentalist. Fecundity has to be measured by isolating gravid females at the end of the experiment and then counting and weighing the eggs that are deposited (Wise 1975). In contrast to *N. radiata*, some orb weavers deposit their egg sacs in the web, frequently as a string of sacs generated over several weeks or months. David Spiller and I have exploited this feature to test for food limitation in orb-weaving spiders. In an experiment with populations in a mixed deciduous–pine forest in Maryland, adding fruit flies at a rate equal to at least 50% of the natural feeding rate approximately doubled the fecundities of the basilica spider *Mecynogea lemniscata* and the labyrinth spider *Metepeira labyrinthea* (Wise 1979). In an experiment conducted three years later in the same area, supplementing the prey of immature labyrinth spiders increased their growth rate over a 3.5-week period by 60% (Wise 1983). Mature females of two other orb-weaving species responded in a similar fashion in an experiment conducted in a California salt marsh (Spiller 1984a). Adding insects to their webs increased the fecundity of *Metepeira grinnelli* by 80%, and that of the smaller *Cyclosa turbinata* by almost 30%. This difference in response suggests that prey shortages are more severe for *Metepeira*, probably because in this habitat large insects are less abundant than smaller species (D. Spiller, pers. comm.).

Rypstra (1985) has provided experimental evidence that the orb weaver *Nephila clavipes* aggregates in areas of high prey density in a subtropical Peruvian forest. Before performing the experiment she found that prey activity near aggregations of from three to six spiders was approximately two times that near solitary *N. clavipes* webs. Individuals in aggregations captured prey at a higher rate. In one experiment Rypstra removed prey from the webs of each spider in a group of six spiders for 10 days, with the result that four spiders emigrated. After an additional 10 days, during which no prey was removed, two of the original spiders, plus a new one, had returned. In a second experiment an artificial aggregation of three spiders was created in an area with low prey density. During 10 days of prey supplementation none left and one spider joined the group. All but one of the spiders had left within 10 days after prey supplementation had ceased. Although the number of spiders in the experiments is small, and neither experiment had contemporaneous spatial controls, Rypstra alternated treatments on the same site. The results of both experiments are consistent with the hypothesis that prey availability affects web-site tenacity of *N. clavipes*. This species may frequently experience food shortages. Vollrath (1988) compared growth rates of laboratory-reared *N. clavipes* with those at two different Panamanian field sites, forest edge and forest interior. The daily growth rate of spiders at the edge of the forest was 3 × that of spiders in the interior, but was lower than that of spiders reared in the laboratory on *Drosophila ad libitum*.

Olive (1982) examined the influence of prey availability upon the foraging behavior of a large old-field orb weaver in large screened enclosures built around natural vegetation. Thirty immature *Argiope trifasciata* were added to each of two previously fumigated 4 × 6 m cages that had been divided into four 1 × 2 m quadrats. All spiders in a quadrat received one of four prey rations (coded as 1, 20, 70, 150), in which the lowest had the joules of prey calculated to supply minimum require-

Table 2.1. *Changes in numbers of the orb weaver* Argiope trifasciata *per quadrat in response to prey availability. Ration level is a multiple of the estimated minimum needed for maintenance*

	Number of spiders							
	Cage 1				Cage 2			
	Quadrat				Quadrat			
	A	B	C	D	E	F	G	H
Ration level	1	20	70	150	1	20	70	150
Initial no. (Day 1)	5	6	6	9	3	7	6	5
Final no. (Day 10)	1	9	6	9	2	12	9	5
Change:	−	+	0	0	−	+	+	0
New ration level	150	70	20	1	150	1	20	70
Initial no. (Day 11)	2	10	6	9	2	11	9	5
Final no. (Day 20)	4	14	3	3	4	9	7	6
Change:	+	+	−	−	+	−	−	+

Source: Olive 1982.

ments without growth. Rations were given daily by adding different-sized crickets to the web until the spider had captured the allotted ration. One set of feeding regimes was followed for 10 days for the two replicate cages, at which time feeding treatments were switched among quadrats and the experiment was continued for another 10 days. Spiders tended to abandon sites with the lower rates of prey supply, so that after 10 days spiders were aggregated in the quadrats receiving the higher prey rations (Table 2.1). This experiment clearly demonstrates that the frequency with which *A. trifasciata* re-locates its web is at least partly a function of prey availability, but sheds only indirect light on whether or not this behavior occurs frequently in populations exposed to natural prey abundance and diversity.

In a short-term field experiment with a non-caged population of *Argiope keyserling*, Bradley (pers. comm.) demonstrated that this common Australian relative of *A. trifasciata* responds to differences in feeding rate in a similar manner. Bradley located 80 adult females and fed half of them for four days. Supplementing prey, even for this short period, decreased the tendency to change web site (25% and 39% for fed and unfed, respectively; $p(G) < 0.001$). The difference persisted for a five-day period after food supplementation ceased (14% and 38%; $p(G) < 0.001$).

Rubenstein (1987) found a positive correlation between the rate at which the orb weaver *Meta segmentata* captures prey and the number of conspecific webs within one meter. This pattern suggested to Rubenstein that spiders were aggregating at web sites where prey was most abundant. He tested his hypothesis by first screening webs in good sites and removing natural prey, and then adding 3 or 10 *Drosophila* per hour to the enclosure. At the low feeding rate, 9 of 10 spiders left their webs and moved about the enclosure. No spiders (out of 10) left their webs at the high experimental feeding rate.

The filmy dome spider, *Neriene radiata*, distinguishes between web sites that differ in prey supply, a discovery made by Martyniuk (1983) in an ingenious field experiment. Martyniuk located 36 abandoned and 36 long-term occupied web sites, for which he measured temperature, humidity and prey abundance over a four-week period. Sites did not differ in physical characteristics, but the occupied web sites had *c.* 50% more prey as determined by sampling with sticky traps. Martyniuk induced 54 field-collected immature filmy dome spiders to construct webs on movable, artificial web sites in the laboratory. He then placed a spider and its artificial web support at each of the 36 abandoned sites, and introduced 18 to half of the occupied sites after having removed the original resident. Half of the spiders that had been introduced to abandoned sites were then given one fruit fly daily for a month. Supplementing natural prey levels increased web-site tenacity. Spiders placed in abandoned sites remained 10 ± 3 days at natural prey levels, but stayed 25 ± 3 days before leaving if given extra prey. This latter residency time did not differ from that of (1) spiders left undisturbed at the long-term occupied sites or (2) spiders introduced on artificial web supports to the long-term occupied sites. Thus an additional fly a day turned a poor site into a good one, which agrees with Martyniuk's original estimate that prey availabilities at abandoned and long-term sites differed by about one prey item per day.

In contrast to Martyniuk's findings, my supplementation of the prey intake of immature filmy dome spiders did not affect web-site tenacity (Wise 1975). I established experimental populations on ground juniper bushes (cf. Chap. 5), which likely provide web sites that are more homogeneous than the array of sites over the entire forest floor. The strength of Martyniuk's experimental design is his prior identification of web sites that appeared to differ in suitability as defined by the spider's behavior, and the subsequent direct experimental test of the hypothesis that the apparent difference in residence times reflected different prey availabilities.

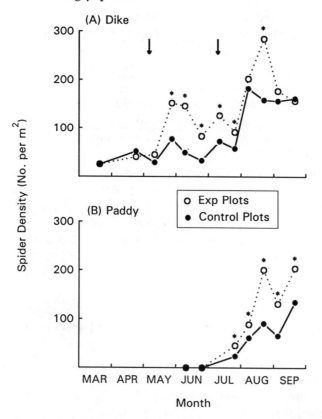

Fig. 2.4 Change in density of spiders in response to release of *Drosophila* on dikes in a rice field. Dotted lines indicate seasonal change in the mean of the two plots on which fruit flies were released; solid lines indicate changes in the mean of the two control plots. Asterisks (★) indicate dates on which there were significant differences (details of statistical test not specified) between control and prey-supplemented plots. (A) Spider densities on the dike in the center of each plot. Arrows indicate the dates of *Drosophila* release. (B) Spider densities in the paddy, i.e. the portion of the plot planted in rice. (After Kobayashi 1975.)

Kobayashi (1975) released *Drosophila* on 30 × 0.6 m dikes surrounded by rice paddies in Japan to determine whether the increased prey density would result in increased spider densities under natural conditions. His manipulation apparently produced a rapid increase in densities of spiders in the dikes and adjacent paddies (Fig. 2.4). Kobayashi employed two experimental (prey-release) and two control areas. Unfortunately, two aspects of the design and analysis make it difficult to assess the observed response: (1) experimental and control plots were not physically interspersed; in fact, the two release plots were immediately adjacent; and (2) mean densities are presented and statistically significant differences are indicated, but no mention is made of the extent of variation between replicate plots or degrees of freedom in the statistical analysis. It is not clear if variation between within-plot samples was used to estimate experimental error; if such were the case, the analysis would have an inflated number of degrees of freedom because of sacrificial pseudoreplication (Hurlbert 1984). However, comparison of the mean densities of spiders after the release with before-release values suggests that a numerical response by spiders did occur.

Are spiders really hungry?

Not all ecologists have concluded that spiders are food limited. Kessler (1971) speculated that '. . . there is always a surplus of food for spiders of the genus *Pardosa* in their habitats,' although he presented no evidence to support this conclusion. Greenstone (1978) hypothesized that '. . . numerical responses to prey availability will tend not to be seen in spider populations in nature because spiders generally find themselves in situations of prey abundance.' He suggested that this general lack of food limitation results from a 'suite of adaptations associated with their phylogeny and life style,' especially physiological adaptations that enable them to withstand long periods of starvation. Greenstone proposed this hypothesis before much of the evidence of food limitation in spider populations had accumulated. Although it is reasonable to postulate that food shortages have selected for traits that minimize the impact of prey scarcity upon fitness traits, it does not follow that spiders have escaped resource limitation. The evidence just presented indicates that prey shortages still frequently limit components of spider fitness. Nevertheless, not all of the evidence suggests food limitation. Some investigators have failed to find correlations between prey density and spider parameters that would suggest food limitation, and a few field experiments have failed to uncover food limitation.

Greenstone (1978) found no correlation between the population density or individual size of the wolf spider *Pardosa ramulosa* and prey availability at 24 brackish pools around which the spiders aggregate. Kajak (1965) found no correlation between the abundance of Diptera in sweep-net samples and the densities of two species of *Araneus* in a meadow, nor did she find a correlation between spider density and the rate of prey capture per spider.

Conley (1985) failed to find evidence of food limitation in a field experiment with the burrowing wolf spider *Geolycosa rafaelana*. Every week from May through October she opened the burrow of all adult females in 20 × 20 m plots, placed a *Tenebrio* larva in the mouth of the burrow and then closed it. Increasing the food supply did not improve survival over the season. Although the plots were large, sample sizes were small because initial spider densities ranged from 6 to 11 per plot, and by the end of the season only one individual per plot remained on experimentals and controls. The manipulation clearly did not improve female survival, but the possibility of food limitation in this species has not been ruled out. Since adult spiders do not re-locate their burrows, food shortages would not be expected to affect emigration. *G. rafaelana* does not forage far from the burrow mouth (Conley 1985); therefore, the most likely effect of food shortages on survival would be increased mortality resulting from increased time at the mouth of the burrow waiting for prey. It is unknown whether supplementing prey decreased the foraging time of *G. rafaelana*. It is more likely that a shortage of prey would limit fecundity of *G. rafaelana*. However, because eggs are deposited in the burrows, fecundity cannot be measured without severely disrupting the entire system (Conley 1985). Possibly food availability plays a role in locating the first burrow by dispersing young spiders. No results from field experiments are available on this question, but in a laboratory study with two other species of *Geolycosa*, Miller (1984) found that dispersing spiders established burrows with greater frequency if they had recently fed.

Schaefer (1981) compared the standing crop and prey consumption rate of a linyphiid and two wandering spiders with the biomass of prey in the environment, and concluded that these three species were not food limited because they captured a relatively small fraction of their prey populations. However, ratios of standing crops may be misleading, because it is the rate of supply of the resource that determines availability. Supplementation of the natural rate of supply is the most direct test of whether this rate is limiting. Schaefer (1978) performed such a field experiment with the web builder of the above study and found no evidence of food limitation. This evidence against food limitation is also not conclusive. Rather than directly increasing the rate of supply of prey to each web, Schaefer placed open culture bottles of fruit flies in the natural habitat; it is not known whether or not significant numbers of the introduced flies actually reached the webs and were captured by the spiders.

Enders (1975a) found that increasing or decreasing prey availability did not affect web-site tenacity of the orb weaver *Argiope aurantia*; however, the two periods of prey manipulation in his field experiment lasted only four and six days each. His conclusion that prey availability does not affect the web-site tenacity of *A. aurantia* does not agree with Olive's (1982) longer study with *A. trifasciata* that was conducted, however, under semi-artificial conditions.

Failure of prey supplementation to increase web-site tenacity is not proof that food is an unlimited resource. A spider may continue to experience prey shortages even though levels are high enough to prevent its abandoning a site. For example, in the population of *Neriene radiata* that I studied, supplementing prey increased the growth rate of juveniles but did not lower the disappearance rate of spiders from the experimental bushes (Wise 1975). Experiments with orb-weaving spiders (Araneidae) also have shown that a shortage of prey can limit growth or fecundity even though adding prey does not decrease the rate at which web sites are abandoned (Wise 1979, 1983, Spiller 1984a, Spiller and Schoener 1990a).

In my prey-supplementation experiment with the basilica and labyrinth spiders, numbers of spiders unexpectedly declined faster in the group receiving extra prey (Wise 1979). This decline was limited to the first part of the experiment and was more marked for the labyrinth spider, which has a more flimsy web. This pattern led me to suggest that the initially higher rate of disappearance from the supplemented population resulted from excessive disturbance to the spiders as we perfected techniques of adding flies to the webs. Caraco & Gillespie's (1986) model predicts that an increase in prey availability can cause a spider to abandon the sit-and-wait strategy, which tempts me to re-evaluate my unexpected results. However, claiming that they confirm the predictions of Gillespie & Caraco would be unwarranted, because their prediction depends upon a high temporal variability in prey abundance. If temporal variability is low enough that spiders can quickly assess the quality of a web site, spiders should increase web-site tenacity in response to an increase in prey supply (Gillespie & Caraco 1987). Thus the appropriate experimental test of their model would be one that varied average prey abundance and its variance over space and time.

The above brief digression illustrates the difficulties inherent in analyzing results of studies as simple and direct as supplementing the prey of predators in their natural surroundings. Accumulating empirical results and predictions of foraging models suggest that understanding

how web-building spiders respond to prey abundance is more complicated than originally envisaged.

Synthesis

The degree of food limitation varies spatially and temporally, and factors
other than prey abundance can also influence web-site selection, growth
and survival. Nevertheless, the pattern that emerges clearly leads to the
conclusion that food often is a limited resource for spiders. Several
observational studies suggest that web-building and wandering spiders
frequently face a shortage of prey in nature. They often do not achieve
maximum possible rates of growth and reproduction and in some cases
may suffer lower survival because of food shortages. This latter effect
may be most pronounced among spiderlings, though this is only a guess
because the earliest stages have not been adequately studied. Among
older spiders effects of food scarcity on survival may be more indirect,
such as mortality resulting from increased exposure to predation as web
sites are abandoned in search of more productive ones. Field experiments
indicate that food often is a limited resource for spiders.

Some studies have uncovered numerical responses by spider populations to increases in prey numbers. However, much of the evidence,
including results of field experiments, is largely restricted to demonstrating within-generation effects on growth and fecundity. More research is
needed on between-generation influences of prey limitation. It is clear
that spiders are hungry most of the time; we now need to know the
extent to which this hunger translates into a reduction in population
density. Other limiting factors may compensate partially for increases in
survival and fecundity that result when prey abundance is increased over
ambient levels; whether such compensation frequently occurs, and its
magnitude, would be fruitful areas of future research.

Synopsis

A predator is food limited if a shortage of prey limits population density.
Although the concept of food limitation implies a population-level
response measured over more than one generation, evidence of food
limitation in spiders most frequently comes from within-generation
responses to changes in food availability, i.e. from effects of prey
shortages on foraging behavior, individual growth, survival and reproduction. Food contains more than joules, and possibly food limitation of

spider populations involves more than a shortage of energy. Most commonly, though, spiders are calorie-limited; as broadly polyphagous predators, they are more likely to suffer a relative shortage of joules than a shortage of particular amino acids or other nutrients.

Several attributes of spiders are consistent with the hypothesis that a shortage of prey has been a major selective factor over evolutionary time. Spiders can survive long periods of starvation, primarily by waiting for prey rather than by actively searching for it, and also by lowering their basal metabolic rate in the absence of enough prey to support growth and reproduction. Foraging patterns appear to have evolved under pressure from food limitation. For example, phenologies of orb weavers often coincide with the availability of insect prey in a way that appears to maximize energy input throughout the life cycle. Within a season, spatial foraging patterns of spiders appear to have been molded by prey shortages. It has been proposed that the sit-and-wait strategy of many spiders is an adaptation to a shortage of prey. When spiders change foraging sites they often re-locate in microhabitats of higher prey abundance.

Evolutionary responses to past prey shortages have not prevented spiders from being hungry over ecological time. Indirect evidence of food limitation includes: extensive variation within populations in growth rate, size at maturity and fecundity; correlations between prey abundance and growth, fecundity and population density; and rates of growth and fecundity under laboratory conditions that are higher than occur in nature.

Controlled field experiments provide the most direct evidence of food limitation in spider populations. Web-building spiders offer two distinct advantages to the experimental ecologist: their sedentary habits eliminate the need to build restraining barriers, which would alter both prey and natural enemies; and their webs offer a convenient vehicle for supplementing natural feeding rates. Supplementing natural prey in controlled field experiments has uncovered clear evidence that a shortage of food limits growth and reproduction of a linyphiid and several species of araneid orb weavers. Field experiments have also demonstrated that web builders and wandering spiders will aggregate in areas with higher rates of prey supply.

Not all ecologists have concluded that spider populations are food limited. It has been argued that evolved adaptations to food shortages are not evidence of current food limitation, but instead permit spiders to escape food limitation over ecological time. Some studies have failed to uncover correlations between prey abundance and rates of prey capture or spider population density. Field experiments have uncovered no evidence of food limitation in a linyphiid, an orb weaver and a burrowing wolf spider. However, limitations of experimental design in these experiments mean that food limitation cannot be ruled out for these species.

Food is most likely not a limiting factor for all spider populations in all years or habitats. However, accumulating evidence, both indirect and direct, makes it clear that spiders are frequently hungry, to the point of exhibiting rates of growth and reproduction considerably below what is physiologically possible.

3 · *Competitionist views of spider communities*

Early threads

Spiders are generalist predators with overlapping diets who are hungry most of the time. They ought to compete with each other for food. It is not surprising that arachnologists have invoked interspecific competition to explain patterns in spider communities. The original impetus was neither experimental evidence of food limitation nor general arguments, such as that advanced by Hairston, Smith & Slobodkin (1960) that terrestrial carnivores are food limited and therefore should compete. Instead, motivation sprung from the fascination with interspecific competition that has permeated ecology since Charles Darwin. An early, extensive statement of the competitionist view of spider communities occurs in the review by Erwin Tretzel (1955).

Tretzel used the term 'intragenerische Isolation' to describe differences in spatial and temporal distribution observed for congeneric species; the term could be translated broadly as 'niche partitioning within genera.' Espousing a world view held by other ecologists of his time, Tretzel argued that interspecific competition has been a major cause of ecological isolation between closely related spiders, and that evolved differences in seasonal timing of reproduction, horizontal habitat utilization and daily activity patterns are the most common adaptations permitting the coexistence of competing spider species (Fig. 3.1). Vertical stratification in the same area is not as effective, he reasons, because of the lack of prey specialization shown by most spiders. Before presenting numerous examples of spatial and temporal segregation in spiders, Tretzel traces the history of the concept. Referring to his studies of wolf spiders, Dahl wrote in 1906 (cited by Tretzel) that 'es unter den einheimischen Spinnen nicht 2 Arten gibt, welche genau die gleiche Stellung im Haushalte der Natur einnehmen' ('there are no two species of indigenous spiders that occupy precisely the same position in nature's household'). Bristowe (1941) consistently argued for the importance of interspecific competition in influencing spider communities. Tretzel

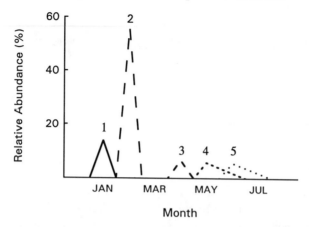

Fig. 3.1   An example of temporal isolation of five spider species, based upon a simplified representation of phenological patterns discussed by Tretzel.
(1) *Typhochraestus digitatus*, (2) *Macrargus carpenteri*, (3) *Trochosa terricola*,
(4) *Tarentula aculeata*, (5) *Xerolycosa (Alopecosa) nemoralis*. The first two species are sheet-web weavers; the others are wolf spiders. (After Tretzel 1955.)

(1955) cites Bristowe's discussion of the fact that the number of spider species in a genus in the Scilly islands declines with island size: 'This same circumstance can be noted for each set of species with similar habits competing for the same set of insects. Although we must not blind ourselves to climatic and other factors, it is probably fair to assume in a general way that those species which we find on one of these small islands have survived a struggle for existence with other species of similar size and habit.' Tretzel believes that competition for food and space has led to spatial and temporal 'Isolation' among spiders, a belief he shares with Bristowe (1941): 'Different species have adapted themselves to different habitats and within each habitat there are spiders to attack large insects and others to capture smaller kinds, spiders to attack diurnal insects and others which hunt by night, spiders which are a danger to crawling insects and others which specialize on those which fly. Modified structure and modified instincts combine to make the greatest use of the available food supply' (cited by Tretzel 1955).

Tretzel points out that certainly not all differences between similar species result from competition. Nevertheless, after examining the patterns exhibited by spiders he concludes that '. . . man kann die Bedeutung der Konkurrenz als Ursache der intragenerischen Isolation nicht einfach damit ablehnen, daß man die Sonderungserscheinungen "nur" also Folge von Artbildungsvorgängen ansieht. Das heiße eine

Wirkung ohne Ursache annehmen zu wollen' ('one cannot easily reject the meaning of competition as the cause of "intrageneric separation", one cannot regard the phenomena of separation "only" as resulting from speciation events. That would be willingness to accept an effect without a cause').

Focus on interspecific competition among spiders has continued during the decades since Tretzel's assertion of its importance in explaining spatial and temporal distributions. Several studies have explicitly inferred the operation of competition from the patterns they uncovered, whereas others have documented niche differences among spiders against the broad interpretive background of competition's driving force. The competitionist world view even flows implicitly beneath the surface of studies whose central theme is not competition itself. Below is a brief sampling of some of the more oblique overtures to competition, followed by an examination of the indirect evidence for interspecific competition among spiders that has accumulated since Tretzel proposed its importance.

Allusions to competition among spiders

Studies that refer tangentially or implicitly to past or current competitive pressures cover the gamut from population ecology to behavior and sensory physiology. Bleckmann & Rovner (1984), in a study relating the micro-distribution of the semi-aquatic wandering spider *Dolomedes triton* (Pisauridae; closely related to Lycosidae) to its ability to sense prey on the water surface, speculate that '. . . *D. triton* is probably exploiting food resources unavailable to spiders that are not adapted for a semi-aquatic life.' Craig, Okubo & Andreasen (1985) studied the impact of web oscillations on prey capture, finding for example that spiders that feed on smaller prey build webs that oscillate more. This discovery led the researchers to conclude that '. . . web oscillations are important to determining ecological differences, among adult spiders of different sizes, as well as among spiders of the same species but different age classes.' Although in neither publication do the authors make explicit mention of competition, what is the significance of ecological differences in prey capture if not based in response to, and probable avoidance of, interspecific competition?

Nyffeler & Benz (1979a) compared overlap in spatial and prey niches of crab and wolf spiders in mown and unmown fields. In mown fields the two families forage in the same area, but differences in prey are still substantial. Predation by crab spiders in the unmown fields may be an important source of mortality for lycosids. Nyffeler & Benz do not mention the possible role of competition between these families, but assumptions about its potential importance likely sparked the research. Several other studies of niche partitioning and overlap between spiders also do not focus on competition, but may have been motivated by pervasive assumptions about competition's importance in ecological communities (e.g. Kuenzler 1958, Vermeulen & Kessler 1980, Kessler, Vermeulen & Wapenaar 1984, Pasquet 1984a,c, Nentwig 1985b).

Direct references to interspecific competition among spiders are much more frequent than oblique allusions to its role in shaping ecological phenomena. Several researchers have speculated on the importance of interspecific competition even though it was not a primary focus of their research. For example, in a survey of spiders inhabiting an oak stand in Wytham Wood, England, Turnbull (1960a) asserts that some species, primarily web-spinning spiders, leave the ground layer in spring and early summer in order to escape high levels of competition that result from high spider densities. Toft (1978) states that '. . . adaptation to life in a certain stratum implies conformity to certain phenological patterns, the advantages of which must be larger than the drawbacks resulting from increased competition.' Several ecologists (Huhta & Viramo 1979, Aitchison 1987, Kirchner 1987) have suggested that some spider species have responded to competitive pressures from other spiders and non-arachnid invertebrate predators by becoming active on, or under, the snow in winter. If this is true, these species would be the arachnid equivalents of the Inhalmiut, the inland Eskimos of the Barrenlands, who may have been driven into their frigid, forlorn niches by more aggressive competitors (Mowat 1975). Aitchison (1987) believes that this evolutionary exodus may not yet be complete, as '. . . probably the niche of winter-invertebrate predation is not filled fully in northern climates.'

Duffey (1975), in describing a study of habitat selection by spiders in perturbed, man-made environments, states that species that colonize experimental swards and that also occur in a wide range of biotopes '. . . obviously have a higher competitive ability.' Łuczak (1979) believes that the most abundant species in a community 'are the species winning in the competition with other species,' a view of community organization which leads her to conclude that some spiders are more abundant in crop fields than in more natural ecosystems because of reduced competition caused by lower numbers of spiders in agricultural fields. Nørgaard (1951), after demonstrating a correlation between the microhabitat

distribution of two lycosid species and temperature/humidity preferences and survival in the laboratory, speculates that '. . . as both species
are abundant and feed upon similar small animals, they might no doubt
compete with each other, and eventually also prey upon each other, if
they both lived in the same habitat. Probably these biotic factors work
together with the aforesaid physical factors in causing the difference in
distribution of the two species.' The question of reproductive isolation
motivated den Hollander & Lof (1972) to determine the extent of habitat
separation between two *Pardosa* species; at the end of their discussion
they speculate that the slight differences in habitat utilization they found
'may be caused by interspecific competition.' In a report on the feeding
ecology of a tropical spitting spider (Scytotidae) and a morphologically
similar species in another family (Pholcidae), Nentwig (1985a) states that
'. . . both species compete successfully with coexisting spider species.' He
further proposes that these two species may exploit the same food
resources, yet coexist through temporal (diurnal) niche separation. The
scytotid spitter is a specialist from the tropics, where most spider prey-
specialists occur (Riechert & Łuczak 1982). Riechert & Łuczak argue that
prey specialization in spiders has evolved in the context of competition
for food, and that '. . . through specialization, various spider species are
avoiding competition with the more successful feeding generalists.'

The preceding studies substantiate the importance of competition in
the minds of spider ecologists, but not its prevalence in nature. However,
many studies that have focused explicitly upon interspecific competition
do provide indirect evidence of its importance in spider communities.

Indirect evidence of competition among spiders

Indirect evidence for competition among spiders, as with most organisms, comes basically from two types of research. In the tradition of
Tretzel, the first type is careful documentation of patterns in niche
parameters that are consistent with the hypothesis of interspecific
competition. This approach, though occasionally limited to a species
pair, usually explores patterns within a spider community or guild.
Competition is invoked either as a selective factor acting over evolutionary time – strong but intermittent, or weak but consistent – or as a major
interaction over ecological time, i.e. one or a few generations. In some
interpretations no clear distinction is made between evolutionary and
ecological time scales. The second approach is to uncover a 'natural
experiment,' which is not really an experiment at all. One locates non-

manipulated areas that differ, usually for unknown reasons, in the abundance of an hypothesized competitor. One then determines whether parameters in populations of the other species differ in the direction expected if competitive release is occurring in areas where the putative competitor is absent or less abundant.

Inferences derived from niche variation

In late summer and early autumn, species of large orb weavers mature, becoming conspicuous even to the casual naturalist. Two brightly colored North American species, the araneids *Argiope aurantia* and *Argiope trifasciata*, inhabit grassy and herbaceous vegetation ranging from old fields and prairies to backyard gardens. Similarities in size, phenology, web structure and habitat distribution have made this species pair a natural subject for a range of studies in the competitionist mold. Enders (1974) found that as juveniles, *A. aurantia* built its web lower in a stand of *Lezpedeza* than did *A. trifasciata*, but the difference disappeared by the end of the growing season. Enders interprets vertical stratification between the earlier instars as resulting from competitive pressures. He argues that *A. trifasciata* has evolved into a spatial niche that reduces competition from the evolutionarily older *A. aurantia*.

Brown (1981) used discriminant analysis and multiple linear regression to investigate niche partitioning among these two *Argiope* species and a third species, *Araneus trifolium*, in different sites in two regions of central North America. The discriminant function that explained almost 80% of the variation in foraging patterns of adult females was most closely associated with web height. *Araneus* tended to build its web the highest, and *A. aurantia* the lowest in the vegetation. Webs of *A. trifasciata* were intermediate. Based upon differences in web site, prey size and prey taxa found in his study as well as in several others, Brown concludes that '. . . quite a bit of inferential data indicates that interspecific competition may occur among orb-weavers.' Brown points out that apparently greater divergence in foraging patterns in areas that are drier, and which may thus have lower availabilities of prey, constitutes additional evidence for interspecific competition in the communities he studied. Brown suggests that differences in web height likely cause differences in the size and taxa of prey captured, which would lessen exploitative competition for food.

In contrast to Brown's finding of clear differences in web height between adult *A. aurantia* and *A. trifasciata*, and in agreement with

Enders' (1974) results, Uetz, Johnson & Schemske (1978) found wide overlap in vertical placement of the web. However, they did find significant differences between the two species in the size of opening in the vegetation utilized for web construction. Uetz *et al.* suggest that differences in both web placement and web mesh (i.e. size of the openings in the catching spiral) modify each species' diet sufficiently to permit coexistence. Differences in prey size and taxonomic composition of the diet were statistically significant. McReynolds & Polis (1987) have also studied niche utilization patterns in this *Argiope* pair, documenting wide overlap in the types of prey utilized. However, they discovered no clearly discernible effect of mesh density on the type of prey captured.

These studies with different populations of the two *Argiope* species yield patterns that often are compatible with a competitionist interpretation, but that are not entirely consistent. Unfortunately, no other comparable pair of web-building species has been subjected to such consistent competitionist scrutiny.

Greenstone (1980) has interpreted the distribution pattern of two wolf spiders in the context of resource partitioning and avoidance of competition. *Pardosa ramulosa* and *P. tuoba* are broadly sympatric through much of coastal California, yet are sharply segregated on a smaller scale: *P. tuoba* occurs only in scrub and prairie habitats, never in the salt- and freshwater marshes inhabited by *P. ramulosa*. When the latter is found outside marshes, it is restricted to the margins of small pools. This distribution correlates well with the relatively specialized feeding behavior of *P. ramulosa*, which preys primarily on insects at the surface film of small bodies of standing water. Greenstone (1980) discusses other examples of distribution patterns within the genus *Pardosa* and concludes that '... *Pardosa* species have been constrained in such a way that many of the mechanisms we associate with avoidance of competition have not been open to them. Rather ... they have tended to become very specific in their habitat requirements.'

Studies of patterns within larger assemblages of species, i.e. spider communities, are more numerous than studies of overlap in the niche parameters within isolated species pairs. Recently such approaches have become more formalized with the development of niche theory and the application of multivariate statistical procedures.

Łuczak (1966) interprets patterns in the relative abundances of wandering spiders in the heather and tree layers of eight different pine stands as resulting from interspecific competition over ecological time scales. Dondale (1977) reasons that differences in phenologies and habitat distributions of wandering spiders within a meadow promote species diversity by minimizing interspecific interactions, implying that competition has been an important evolutionary force. Chew (1961) proposes competition as the factor to explain most spatial and temporal distributions of spiders in a desert community. Gertsch & Riechert (1976) utilized principal components and cluster analysis to categorize a desert community of 90 species into groups within which frequent interaction would be expected. They concluded, on the basis of their data and an analysis of the available literature on spiders, that '. . . in most, but not all cases, closely related species are separated by spatial and temporal differences.' They note that only a few pairs or triplets exhibit no noticeable niche partitioning. Gertsch & Riechert suggest that interspecific competition may be alleviated by intrageneric predation, or by unpredictability of abiotic factors that would cause an unstable community equilibrium. Underlying their interpretation is the assumption that competition has molded the evolution of the community, even though they apparently believe that competition currently is weak or absent.

Fowler & Whitford (1985) utilized similar multivariate techniques to describe patterns in a community of three web-building desert spiders. They write that 'separation in the niche dimensions of time, weather and habitat aspect were found, suggesting that interspecific competition may be important in structuring the community.' Unlike Gertsch & Riechert, Fowler & Whitford conclude that significant competition is occurring over ecological time scales. They attribute a negative correlation between the abundances of two species to direct competitive interactions, adding that results of a laboratory study support this interpretation. Males of one species did affect web construction of the other, but only at elevated laboratory densities, which suggests that the behavior may be an artifact of confinement.

Uetz (1977) analyzed niche relations in a guild of 10 wandering spiders by calculating niche breadths and overlaps (Levins 1968, Cody 1974) across microhabitat and season in a temperate deciduous forest. He discovered patterns consistent with classical competition theory. For example, differences in relative size of the five smaller spiders were of the magnitude proposed by Hutchinson (1959) as sufficient to permit coexistence of co-occurring similar species. Larger species, however, were more similar in size than the critical Hutchinsonian ratio, leading to the prediction that they should segregate along another niche dimension. In agreement with this prediction, the two largest species were either 'habitat or seasonal specialists.' Within the entire guild there was no statistically significant relationship between a species' niche breadth over habitats and its temporal niche breadth. However, Uetz points out that '. . . segregation of seasonal occurrence and/or habitat is apparent in the sub-groups and in pairs of congeners within each group.' A finding in agreement with competition theory is that '. . . overall overlap between potentially competing species based on size or congeneric status is significantly lower than overlap with all other species.' In summary: although no pattern of general niche complementarity in this guild of 10 wandering spiders emerges, one can find evidence of segregation when one compares species most likely to compete, i.e. those most similar in size or phylogenetic relationship. Another interesting outcome of Uetz's study is the finding that temporal overlap is less than spatial overlap, which agrees with Tretzel's (1955) generalization on the impact of competition among closely related spiders.

Suwa (1986) described niche partitioning among four wolf spiders of the genus *Pardosa* in Japan, which he attributed primarily to differences in habitat preference and interspecific competition. He discovered extensive overlap in diets, life-cycle phenologies and daily activity patterns; he calculated that the species' size ratios were less than the criterion that Hutchinson (1959) speculated permits competitive coexistence. On the other hand, Suwa found patterns in spatial overlap that are consistent with predictions of competition theory. Species that show wide geographic overlap occurred in different habitats, whereas species that were found in similar habitats were more geographically segregated. In addition, when distributions overlapped on a broad scale, the species were distributed allotopically on a finer scale, a situation reminiscent of the two *Pardosa* species in California studied by Greenstone (1980).

Post & Riechert (1977) estimated spatial overlap between species from an 11-day series of 0.1 m² samples collected from three contiguous spider communities (two old fields and a deciduous woodlot). The investigators used these data to approximate the interaction coefficients of Levins' (1968) community matrix because they were '. . . interested in possible competitive interactions (or avoidance of such interactions through niche partitioning).' Their approach explicitly assumes that densities of the numerically dominant species are at equilibrium, and that patterns of spatial overlap can be related to interspecific competition when combined with natural history information and the 'assumption that the competitive exclusion principle applies.' They conclude that species that are habitat specialists avoid competition with more generalist species, but often compete with other habitat specialists. Similarly to Uetz (1977), Post & Riechert discovered that a pair of lycosids (*Pirata* spp.) did not differ sufficiently in size to avoid competition according to the criterion of the Hutchinsonian ratio. These species were not spatially segregated; the sampling regime was not designed to uncover temporal separation.

Adult female linyphiids living in Danish *Calluna* heath display Hutchinsonian ratios ranging from approximately 1.10 to 1.25, with the larger species in the sequences tending to exhibit higher ratios (Toft 1980). This pattern characterizes communities of sheet-web weavers from several habitats (Toft, unpubl. data). Toft argues that this pattern is consistent with conventional competition-based theories of limiting similarity if individual niche widths increase with size and/or if the coefficient of variation in body size increases with size; both patterns have been documented for linyphiids (Toft, pers. comm.). Toft further argues that competition for prey or web sites could explain the sequence of body sizes observed in linyphiid communities, since both prey size and web volume are correlated with linyphiid body size.

Natural experiments

Static patterns of variation in niche parameters often are consistent with competition theory, but also have alternative explanations, ensuring that tenacious advocacy of competition as the underlying cause will provoke occasionally animated discussion (e.g. exchanges of viewpoints in Strong *et al.* 1984). Stronger proof of competition's importance would come from evidence of release from competition in communities from which putative competitors have been experimentally removed. Such experiments are not always feasible, which makes the so-called 'natural experiment' attractive. One performs a natural experiment by equating uncontrolled spatial or temporal variation in the density of one or more species with a planned manipulation of the hypothesized competitor. This logistical sleight of hand is accomplished simply by assuming no confounding variation in other potentially influential factors.

Several investigators have concluded that competition caused changes in community structure between years or over a spatial gradient. Łuczak (1963) for example, in a study of web-spinning spiders on the heather and trees of eight pine plantations, found that the most abundant, or 'dominant' species decreased dramatically in numbers from one year to the next concurrently with an increase in abundance of other species in the community. She attributed the increase in numbers of the less common species to release from competition with the dominant species, which she hypothesized decreased the second year because the weather changed. Otto & Svensson (1982) studied the distribution of ground-living spiders along altitudinal gradients, and suggested that many species found at higher elevations 'reflect cases of ecological release'; that is, the high-altitude species are 'fugitive species' that do not do well at lower elevations where competition is severe, but do much better in higher habitats where several lower-elevation species are absent.

Studies such as those of Łuczak (1963) and Otto & Svensson (1982) are natural experiments, though these researchers do not label their research as such. Their approach resembles the explicitly identified natural experiment because they infer the operation of interspecific competition over ecological time by relating shifts in community composition to temporal or spatial changes in densities of hypothesized superior competitors. Similar approaches have been used by Uetz et al. (1978) and Yoshida (1981) to explain patterns among small sets of orb-weaving spiders. Uetz et al. (1978) propose that differences in habitat distribution, in particular the building of webs at different heights and in different-sized openings in the vegetation, facilitate the coexistence of forest-dwelling orb weavers with similarly spaced threads in the catching spiral. As an example they cite the vertical distribution of Mangora placida. This species constructs webs up to 1.2 m above the ground in a forest in which a species with similar mesh size, Metepeira labyrinthea, is absent. In another forest, in which Metepeira lives and locates webs from 0.5 m to 2 m, Mangora's webs are not found over 0.75 m. At the shared heights in this forest spatial overlap is minimized further because Mangora tends to construct webs in larger openings than Metepeira. Yoshida (1981) examined the distribution along a stream of three orb weavers in the genus Tetragnatha. He does not attribute differences in spatial distribution to competition, but does suggest that web heights may be influenced by the density of other species in regions of spatial overlap. The proposed mechanism is increased frequency of web invasions at higher densities. The data presented are not extensive enough to support any but the most tentative conclusions concerning competition; Yoshida recognizes the limited nature of his small sample sizes and the further troubling fact that '. . . differences of the web height may be due to the original difference of each species.'

Several comparisons have been published that qualify as explicit natural experiments. Two studies involve *Argyrodes* species, which are theridiids that feed on prey in the webs of other spiders, usually from other families. Early natural history literature described these theridiids as commensals, invaders who remove small prey morsels that are of no real value to the host. However, more-detailed examinations revealed that *Argyrodes* often take prey that the host would have eaten and thus are really kleptoparasites (Wiehle 1928 cited by Kaston 1965, Robinson & Robinson 1973, Vollrath 1979). Rypstra (1981) found a negative correlation between the rate of prey consumption by the large tropical orb weaver *Nephila clavipes* and the number of smaller *Argyrodes* in the web. Host spiders with the larger kleptoparasitic load re-located their web at a higher frequency. Although kleptoparasite and host show dietary overlap, the kleptoparasitic interaction is highly specialized behavior in which only one member of the pair is harmed, which disqualifies it from being considered competition as usually defined. The potential does exist for competition between different kleptoparasitic species living in the same host web. Results of a natural experiment described by Vollrath (1976) suggest that *Argyrodes* species compete with each other in the webs of their hosts.

A. *elevatus* and A. *caudatus* inhabit the rain forest on Barro Colorado Island in Panama. During the rainy season both species were equally abundant in the same webs of *N. clavipes*. A. *elevatus* was active primarily during daylight, whereas A. *caudatus* was nocturnal. During the dry season Vollrath (1976) examined *N. clavipes* webs at a site in a secondary growth forest. In this more exposed habitat A. *caudatus* was rare. *N. clavipes* webs were inhabited almost exclusively by A. *elevatus*, which was active only nocturnally, opposite to its behavior in the rain forest. Vollrath postulates that A. *elevatus* was able to shift its activity to nighttime because of the absence of interference from A. *caudatus*. Vollrath (1987) suggests that A. *caudatus* was absent at this site during the dry season because it is not well adapted to withstanding heat stress in open, sunny habitats. The hypothesis of competition leads to two predictions: (1) experimentally removing A. *caudatus* from webs in the rain forest during the rainy season will cause A. *elevatus* to shift its activity to nighttime; (2) introducing A. *caudatus* into webs at the secondary growth site during the dry season will cause A. *elevatus* to become day-active. The second experiment might not be successful because a good chance exists that introduced A. *caudatus* will not persist under the physical conditions of the site in the secondary growth forest. Despite the risks, it would be worthwhile to attempt these manipulations in order to test directly the interpretation of this natural experiment, i.e. that the activity pattern of *A. elevatus* is determined by the intensity of interference competition with the congeneric species.

A relatively recent continental invasion provides the basis for a natural experiment on interspecific competition between two more conventional, non-thieving theridiids. Nyffeler, Dondale & Redner (1986) argue that a European immigrant, *Steatoda bipunctata*, has been displacing its North American relative, *S. borealis*, from microhabitats in the center of cities and towns. Considerable evidence suggests that *S. bipunctata* recently arrived in North America (the earliest museum record is 1913) and has been rapidly spreading through New England, the maritime provinces and along the St Lawrence River. As evidence of competitive displacement Nyffeler and his colleagues point to the following pattern: (1) Only *S. bipunctata* is found in the center of Ottawa-Hull, where it occurs in buildings and on tree trunks. (2) At the periphery of the city either both species occur together or *S. borealis* occurs alone; they have been collected from tree trunks and limestone quarries. (3) Farther from Ottawa-Hull the pattern is mixed on farms that were sampled; some had *S. bipunctata* only, some only *S. borealis*, and others a mixture of both species. (4) On Cape Breton Island, five barns or cottages were occupied only by *S. bipunctata*; two others had *S. borealis* only. (5) All 46 specimens of *S. bipunctata* collected from Maine since 1980 have been found in buildings in cities or towns; the 13 specimens of *S. borealis* came from forests, wooden fences or bridges. Nyffeler *et al.* argue that as *S. bipunctata* spreads from points of introduction in the east, it is displacing *S. borealis* from microhabitats influenced heavily by humans to less disturbed areas.

The pattern is certainly suggestive of interference competition, but problems with this interpretation linger. First, there are no data available that indicate whether or not *S. borealis* has disappeared from particular sites as its cousin has appeared on the scene; i.e. there is no evidence of active displacement. A partial rejoinder to this criticism is a collection made in 1912 in Nova Scotia of 'home-dwelling' spiders that did include *S. borealis*. Secondly, in laboratory tests of aggression, *S. bipunctata* did not displace *S. borealis* from its web, even in staged encounters. However, *S. bipunctata* did not do well in the laboratory; females offered houseflies and fruit flies fed poorly and became progressively shrunken and dull in luster. Thus the crucial behavioral test has not yet been performed. Finally, even though *S. bipunctata* may be spreading and the two species are not always found together, it does not necessarily follow

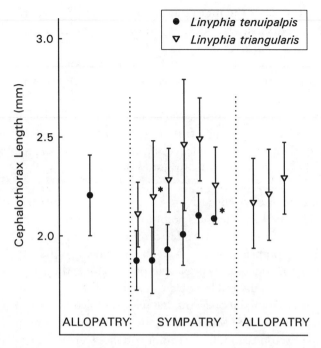

Fig. 3.2 Character displacement between two linyphiid species inhabiting *Calluna* heaths in Denmark. Mean cephalothorax length ±1 s.d. Sample sizes ranged from 18 to 89, with the exception of the two sympatric populations marked with an asterisk (★); these values were seven for *L. triangularis* and three for *L. tenuipalpis*. (After Toft 1980.)

that *S. bipunctata* is displacing *S. borealis* on a local scale. Reasonable alternative hypotheses to competition exist. For example, *S. bipunctata* may be spreading primarily via transport by humans, and may be concentrated in cities or near human habitation because of low rates of dispersal away from these points of introduction. Differences in micro-habitat selection would reinforce the pattern originally dictated by the mode of dispersal. Both competition and non–competition hypotheses are tenable without more data.

Natural experiments have been identified that attempt to explain patterns over evolutionary time. Toft (1980) cites spatial variation in body size of two similar linyphiids as evidence of character displacement driven by interspecific competition. Adults in allopatric populations of *Linyphia triangularis* and *L. tenuipalpis* do not differ in size, yet *L. tenuipalpis* is significantly smaller in sympatry (Fig. 3.2). This divergence is consistent with interspecific competition, and Toft further assumes that the character displacement reflects genetic differentiation. Apparently the size difference in sympatry may be insufficient to alleviate competition completely, because the smaller *L. tenuipalpis* reproduces three weeks earlier than *L. triangularis*. More rapid development of the smaller species reduces the potential size differential, which would intensify interspecific competition. Hutchinson (1959) predicted the opposite phenological pattern for the smaller species of a pair. Ardent competitionists need not despair, as competition theory can still be retained for this linyphiid pair because the species are spatially segregated, '. . . though in a very gradual manner. *L. triangularis* is the dominant species in rich *Calluna*, whereas *L. tenuipalpis* is the dominant in patches of low, poor heather' (Toft 1980).

Eberhard (1989) discovered that the theridiosomatid *Wendilgarda galapagensis*, which is endemic to an island in the Pacific, exhibits a greater diversity of web-building behavior than closely related mainland species. Eberhard suggests that this niche expansion may have resulted from reduced competition with other web-building spiders. This is an intriguing hypothesis, but one that has too many alternative explanations to constitute strong evidence for competition. As Eberhard points out, the niche expansion could also have resulted from decreased predation pressure or increased intraspecific competition on the island.

Synopsis

Spiders are food-limited generalist predators with widely overlapping diets. Many ecologists have assumed, or have concluded on the basis of observed patterns in natural populations, that interspecific competition is a major interaction in spider communities. Tretzel (1955) proposed that interspecific competition accounts for many differences between congeneric species in spatial and temporal distributions ('intragenerische Isolation'). He traced the history of this view and also provided numerous examples. In the years since Tretzel's review interspecific competition among spiders has been widely accepted. Even biologists who have not been investigating competition have alluded to its importance in studies ranging from population ecology to behavior and sensory physiology of spiders.

Ecologists have amassed considerable indirect evidence to support a competitionist view of spider communities. Following directions set by theoretical ecology as well as continuing the approaches advocated by

Tretzel and ecologists studying other organisms, arachnologists have uncovered patterns in niche parameters and species distributions consistent with classical competition theory. Differences in web height, mesh size of the catching spiral and prey characteristics have been interpreted as reflecting competition between pairs of similar orb weavers. The spatial distribution of two congeneric species of wolf spiders suggests they are avoiding competition. Community-wide patterns in niche overlap, especially evidence of complementarity along different niche dimensions, also support the hypothesis of competition within assemblages of both web builders and wandering spiders. This type of evidence often is persuasive, but fails to distinguish clearly between interspecific competition as an evolutionary force and its role as a major interaction structuring communities over ecological time.

Natural experiments provide clearer evidence of the impact of interspecific competition over short time scales. Several comparative studies of spider communities have uncovered apparent shifts in foraging location or spider density that can be attributed to changes in the density of dominant competitors. The two most convincing natural experiments document shifts in foraging behavior or distribution on a local scale in response to changes in density of a congeneric species. There are even two natural experiments that document apparent character displacement over evolutionary time: differences in body size of two linyphiids in allopatry and sympatry, and ecological release in the foraging behavior of an orb weaver that has evolved on an island in the absence of closely related congeneric species.

4 · *Failure of the competitionist paradigm*

The competitionist paradigm

The tradition of explaining community patterns as products of interspecific competition has shaped numerous research programs over the past several decades. Many ecologists have assumed that competition has been a potent selective factor over evolutionary time, and have considered it to be a potentially major interaction in contemporary communities. Interest in competition has been pervasive; it has defined the research programs of theoreticians and shaped the studies of empiricists, who have employed widely accepted approaches to measuring and explaining niche differences and patterns of overlap in nature. The pervasiveness of the focus on competition prompted Strong (1980) to describe competition as a paradigm in Kuhn's (1962) sense of a 'characteristic set of beliefs and preconceptions,' which includes 'instrumental, theoretical, and metaphysical commitments together' (Kuhn 1974). Acceptance of the central importance of interspecific competition was never universal (e.g. Andrewartha & Birch 1954), and recently opposition to the pervasiveness of the concept has increased (e.g. Connell 1975, 1980, Wiens 1977, exchange of views in Strong *et al.* 1984). Simberloff (1982) severely criticized the central role of competition theory as having 'caused a generation of ecologists to waste a monumental amount of time.' Critics have focused upon the lack of attention to alternative hypotheses to competition, particularly in the absence of widespread field experimentation establishing the prevalence of interspecific competition in many natural communities. Criticisms such as Simberloff's are too harsh in light of the accumulating experimental evidence of competition in some communities (Connell 1983, Schoener 1983a), yet it is by no means clear that interspecific competition is pervasive in natural communities. Not all types of ecosystems have been studied with equal intensity, and even for those that have been examined, the species selected for experimental tests of competition have not been random samples of the community.

In a review of the competition controversy, Schoener (1982) concluded that the competitionist view of the world is not a failed paradigm, because '. . . if the results of recent observational and particularly experimental studies can be taken at face value, competition must still be considered of major ecological importance.' Wiens (1983) disagreed, arguing that as a Kuhnian paradigm the competitionist view has failed. In his reply, Schoener (1983b) wondered whether ecology has ever had any Kuhnian paradigms at all, arguing that the competitionist outlook has never been a real paradigm. Although perhaps not strictly Kuhnian in purity, the competitionist outlook in ecology has been so broadly influential that it has served as a paradigm for numerous ecologists, including students of spiders.

The competitionist paradigm has directed numerous studies of spider ecology, but even adherents have expressed caution. For example, Uetz (1977) and Brown (1981) concluded from their research on niche relationships that interspecific competition is important in the spider communities they examined, but tempered their conclusions with the observation that no experimental tests of interspecific competition had yet been made for communities of wandering spiders or orb weavers. Several ecologists, recognizing that the significance of interspecific competition to the functioning of spider communities is questionable in the absence of direct proof, have put the competitionist paradigm to the experimental test.

Absence of interspecific competition in field experiments

Experimentalists have found web weavers particularly attractive. The features that make web spinners well suited for experimental studies of food limitation (Chap. 2) apply equally well to experimental tests of interspecific competition in nature.

Sheet-web weavers

Schaefer (1978) performed the first field experiment to test whether interspecific competition affects the density of a web-building spider. He examined the possible role of several factors (including food supply; Chap. 2) in regulating density of the linyphiid *Floronia bucculenta*, which constructs its sheet web in dense vegetation about 10 cm above the ground. The natural density of *F. bucculenta* was 3 ± 2 females/m² on stands of *Molinia* in a dry *Sphagnum* bog. Schaefer selected open 1 × 1 m

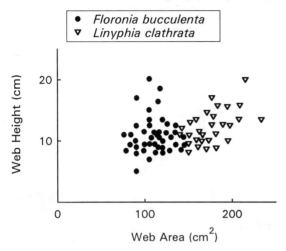

Fig. 4.1 Space partitioning by two co-occurring sheet-web spiders: height and area of the web of the linyphiids *Floronia bucculenta* and *Linyphia clathrata*. (After Schaefer 1978.)

plots for his experimental manipulations. On each of six plots he placed 20 wire cylinders with cross-sectional area equal to that of *F. bucculenta*'s web. He then added 10 *F. bucculenta* to each of these plots, and also to each of five control plots. Suitable web sites were clearly limiting the density of female *F. bucculenta*; mean densities in experimental and control treatments were $10 \pm 2/m^2$ and $3 \pm 2/m^2$, respectively. Addition of 20 *Linyphia clathrata*/m² to another eight plots had no effect on the density of *F. bucculenta*. Also a linyphiid, *L. clathrata* builds its sheet web at the same height as *F. bucculenta*, but in larger openings (Fig. 4.1). These species may avoid competition because they have different web-site requirements.

Schaefer does not indicate whether or not the added *L. clathrata* constructed webs in the experimental plots; if not, the manipulation may not have adequately tested for interspecific competition because (1) a test for exploitative competition for prey between the two species depends upon active *L. clathrata* webs in the plots, and (2) the experiment would not have adequately tested for interference competition if the potential competitor had immediately emigrated. The most direct manipulative approach would have been to remove *L. clathrata* from experimental plots and note changes in growth, reproduction or density of *F. bucculenta*. Competitionists could justifiably argue that Schaefer's experiment does not provide a strong case against the possible impact of interspecific competition on *F. bucculenta*, and thus does not damage the competitionist paradigm.

Orb weavers

Forest spiders

In deciduous forests of eastern North America live two conspicuous araneids that often place their orb webs close together, sometimes even touching, in similar understorey vegetation (Wise 1979, 1981). Both species hang a string of egg sacs conspicuously in the web. The basilica spider, *Mecynogea lemniscata*, is the slightly larger species. Its finely woven orb forms a dome beneath which the spider hangs (Fig. 4.2A). The other member of the pair, the labyrinth spider, *Metepeira labyrinthea*, frequently is found nearby. It hides in a cone-shaped retreat in a maze of silk, connected by a single silken strand to the center of a more typical, almost vertically oriented orb web (Fig. 4.2B).

Basilica and labyrinth spiders build webs in similar vegetation (Wise & Barata 1983; Fig. 4.3). Prey is a limited resource for both species, established by a field experiment conducted in 1977 (Wise 1979; Chap. 2). Thus, these spiders are similarly sized generalist carnivores with similar phenologies, that co-occur syntopically and are clearly food limited. These striking similarities led me to hypothesize that *Mecynogea* and *Metepeira* compete with each other for prey.

In addition to expecting exploitative competition, I predicted that interference competition should occur, based upon indirect evidence obtained earlier in our studies. I had encountered two instances in which a labyrinth spider had spun her web directly over an intact web of a basilica spider. In one instance the labyrinth spider had built her retreat in the top of the basilica spider's dome, and a basilica spider was in the supporting silk above the dome; by the next day she was gone. Two examples were also found in which a basilica female had built her web over an intact *M. labyrinthea* web. Although this evidence was purely circumstantial, the spiders appeared guilty of successful interference competition, in which a web site and a web of another species were the contested resources. These observations suggested a field experiment in which we discovered that both spiders can maneuver in the web of the other and drive the original occupant from the site. Ten webs occupied by mature females of each species were located and a female of the other species was introduced on to the edge of the web. The webs were checked three days later. A labyrinth spider had displaced one of the

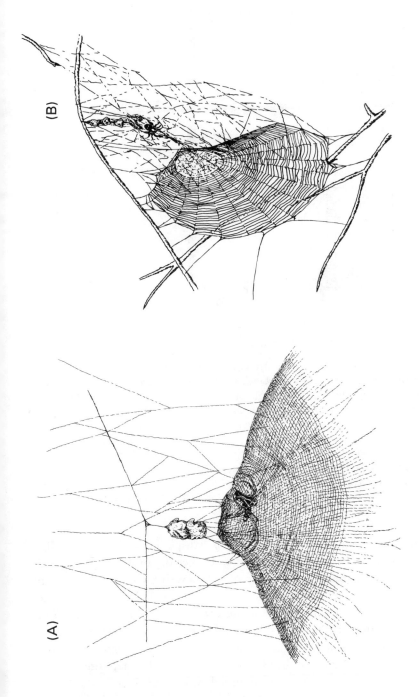

Fig. 4.2 Two conspicuous forest-dwelling araneids: (A) The basilica spider, *Mecynogea lemniscata*, (B) The labyrinth spider, *Metepeira labyrinthea*.

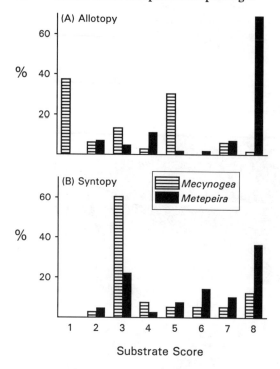

Fig. 4.3 Relative utilization of different types of substrate for web supports by the basilica spider, *Mecynogea lemniscata*, and the labyrinth spider, *Metepeira labyrinthea*. Substrates were scored on a subjective scale of increasing rigidity: (1) live pine, (2) living greenbrier vine, *Smilax*, (3) mixed live and dead greenbrier vines, (4) dead vine, (5) live shrub, (6) live deciduous twigs, (7) mixed live and dead deciduous twigs, (8) dead deciduous twigs. (A) Allotopic populations, several hundred meters apart. (B) Syntopic populations. Substrate utilization overlapped broadly between the two species when they occurred together in the same habitat (syntopy), but *Metepeira* tended to build its webs on more rigid vegetation than did *Mecynogea*. (After Wise & Barata 1983.)

basilica spiders, and five basilica spiders had displaced *M. labyrinthea*. The invaders had constructed their own web over the previously occupied web. Thus 30% of the additions resulted in a web take-over. The bias in favor of the basilica spider likely reflected its larger size at the time the experiment was conducted: *M. lemniscata* was the larger spider in 17 of the 20 encounters. The successful invaders used the old web as support for their own, and in one instance adoption of the conquered infrastructure was so complete that the invader, a basilica spider, deposited an egg sac under the sacs left by the vanquished labyrinth spider.

Thus several lines of evidence led me to hypothesize that the labyrinth and basilica spiders compete with each other. I tested this hypothesis by performing a field experiment in 1978 (Wise 1981), the year after I had demonstrated food limitation in this system.

The basilica and labyrinth spiders build webs in a heterogeneous understorey vegetation at heights from several meters to a fraction of a meter off the ground. Unfortunately for the experimental ecologist, and unlike the situation I discovered for my experiments with the filmy dome spider, webs of this pair were not primarily restricted to small, isolated units of vegetation. Such an arrangement would have been ideal for manipulating population densities. To overcome the limitations imposed by the natural world, I created replicated populations of the two potential competitors on unenclosed units arranged at 10 m intervals throughout a large region of the forest. Each unit was a $4 \times 1.6 \times 1$ m wood frame supporting dead branches of the type both species often use as web supports. The volume of space available for colonization by introduced spiders was $c.$ 10 m^3. Penultimate and recently matured female labyrinth and basilica spiders were collected from the vegetation between the units and from surrounding forests, and were added to the units with conspecifics or in combination with the other species (Table 4.1). Densities on the units encompassed the range of natural densities in the undisturbed population of the surrounding forest. Web positions were mapped on the units, and both disappearances and spiders appearing in new positions were noted three times a week for three months. Survival during the experiment was defined in terms of persistence of a spider in a particular position on the unit. This technique made it possible to exclude at least some immigrants in calculating survival. Spiders were not marked, so that web invasions by conspecific immigrants would have gone undetected. However, this type of interaction would have been scored correctly as an immigration event if the original web resident had moved to a new position on the unit. At the end of the experiment all egg strings were removed and the eggs or spiderlings in each sac (which would have remained in the sac over-winter) were counted.

Neither *Metepeira* nor *Mecynogea* affected the survival or fecundity of the other. The experiment yielded not even a hint that interspecific competition affected these major population parameters. I searched for evidence of competition by defining density of the other species as number of individuals on the unit or as distance to nearest heterospecific. With these indices of density, linear and polynomial regression models of species' effects on survival and fecundity yielded the same result: in mixed-species populations, each species behaved as if the other were

Table 4.1. *Design of the field experiment to test for competition between two orb weavers: the basilica spider* Mecynogea lemniscata *and the labyrinth spider* Metepeira labyrinthea. *Densities are values for Day 0 of the experiment (1 August 1978)*

Treatment	No. of open experimental units	Range in density (no. of females/ unit)	Range in nearest-neighbor distances (m)[a,b]
Mecynogea alone	22	1–13	0.20–2.9
Metepeira alone	18	1–7	0.20–2.8
Mecynogea and *Metepeira*	19	2–19	0.05–2.0[c]
			0.05–2.8[d]

Notes:
[a] Calculated for units with more than one spider. Distance to nearest neighbor, either species.
[b] Median nearest-neighbor distances were 0.6 m for *Mecynogea*, 0.7 m for *Metepeira*.
[c] *Mecynogea*.
[d] *Metepeira*.
Source: Wise 1981.

absent (Wise 1981). Only two probable interspecific web take-overs were detected; in both instances labyrinth spiders had appropriated a *Mecynogea* web site.

An old-field pair

Skeptics might argue that one should look for competition between species that are more similar than the basilica and labyrinth spiders. In retrospect, one might argue that these species are not the best candidates to exhibit intense competition over ecological time. The congeneric orb weavers *Argiope aurantia* and *Argiope trifasciata* are excellent candidates (Fig. 4.4). Their patterns of resource utilization broadly overlap (Muma & Muma 1949, Fitch 1963, Enders 1974, Tolbert 1976, Taub 1977, Uetz *et al.* 1978, Brown 1981). Differences in web structure and placement have led ecologists to suggest that competition between them is an important interaction (Enders 1974, Uetz *et al.* 1978, Brown 1981). Inconsistencies in some of the observed patterns (Chap. 3) could reflect differences in the intensity of competition due to differences between studies in resource levels and population densities of the two species, or the inconsistencies could result from responses to non–competitive

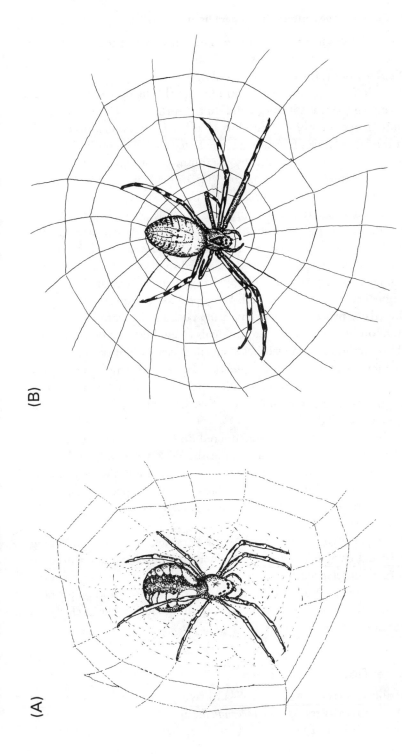

Fig. 4.4 Two orb weavers common in old fields: (A) *Argiope aurantia*, (B) *Argiope trifasciata*.

factors in the environment. Charles Horton and I tested directly for interspecific competition in a set of field experiments conducted over two field seasons (Horton & Wise 1983).

Both *Argiope* species are common in old fields, maturing and depositing egg sacs at summer's end. In the spring of 1979 we divided a field into unenclosed 12 × 12 m plots arranged in two blocks in order to accommodate a moisture gradient. We tested simultaneously for intra- and interspecific competition by establishing two replicates of the following treatments for each species: (1) natural densities of both species together (mixed-species treatment); (2) natural density of one species, density of the other reduced as much as possible – to less than 50% of natural density (single species treatment); (3) high densities (2–3 × natural) of the mixed-species treatment; (4) high density of the single-species treatment. In the first year we added spiders to the high-density plots and removed individuals from the single-species treatment, leaving the original residents undisturbed. We repeated the experiment in 1980, but altered our procedure by first clearing both species from all plots, and then randomly reintroducing individuals to establish the desired densi-ties. This method minimized between-plot variation in spider size and web height. In both years manipulations were performed early in the season and immigrants were continually removed from the single-species treatments. Populations were followed through the end of the reproductive period.

Our density manipulations uncovered no evidence that interspecific competition affects web placement, measured by web height and length of supporting bridge thread; number of prey captured; category of prey captured; individual growth rate or survival. Each species behaved as though the other were not there. Interspecific competition was not detected either year.

Field experiments with two linyphiid species, two forest-dwelling araneids, and two closely related old-field araneids have uncovered no evidence of interspecific competition. Do results from these studies constitute a compelling pattern (Fig. 4.5)? Ecologists would be prudent to use more than one hand in generalizing about interactions as elusive as competition.

An exception to the pattern

The results of David Spiller's (1984a,b) two years of field experiments with the orb weavers *Metepeira grinnelli* and *Cyclosa turbinata* do not conform to the pattern established by the earlier field experiments with

Fig. 4.5   Impatient empirical ecologist extrapolating from results of field experiments on interspecific competition among spiders. (© 1991 by Sidney Harris.)

other species. After the first year of study, Spiller concluded that '. . . interspecific competition was significant and appeared to play an important role in structuring their communities' (Spiller 1984a). In the second year he uncovered evidence of interspecific competition, but concluded that '. . . the ecological significance of the observed processes was ambiguous' (Spiller 1984b). Despite these reservations, he expressed faith in the role of competition when he suggested that '. . . in this system, temporal variability in the competitive abilities may promote species coexistence.' Spiller's studies merit a detailed examination, because their results run counter to those of other field experiments, and because they are well-designed and focus on an intriguing system, one in which differences in phenology influence the symmetry of the competitive interactions.

These two species were the most numerous orb weavers on the edge of a levee bordering a salt marsh in northern California (Spiller 1984a). *C. turbinata*, which matures at a smaller size than *Metepeira*, completes two generations a year, whereas the latter, like its relative *M. labyrinthea*, is an annual species whose spiderlings emerge from the egg sac in early spring and become mature in summer, reproducing in late summer and autumn. A consequence of the different phenologies and differing sizes at maturity is a reversal in relative size over time, with *Cyclosa* being larger than *Metepeira* in spring, and *Metepeira* individuals being larger in summer and autumn (Fig. 4.6A). Another intriguing pattern is the change in relative web height over time (Fig. 4.6B); in early spring *Metepeira* lives much higher in the vegetation than does *Cyclosa*, but by May their web height distributions broadly overlap, primarily because most *Metepeira* shift their webs to lower sites as they mature. A field experiment revealed that both species are food limited in this habitat (Spiller 1984a; Chap. 2).

Spiller (1984a) tested for interspecific competition with a removal experiment. Twice weekly for six months he removed *Metepeira* from three 4 × 1 m plots and *Cyclosa* from three other plots; he left both species on three control plots. The plots were open patches of natural vegetation. The approach was similar to that used during the first year by Horton & Wise (1983). The vegetation structure made it possible for Spiller to isolate and manipulate replicate populations without having to use artificially placed web sites, which I had to use when studying the related *Metepeira labyrinthea* in a forest.

Spiller compared several parameters between control and experimental plots over the entire season by calculating separate *t*-tests. The manipulations produced evidence of interspecific competition in this system. The pattern is complicated, primarily because different competitive mechanisms are involved. Reducing *Cyclosa* densities appeared to reduce effects of exploitative competition for prey on *Metepeira*, based on the following statistically significant higher values for *Metepeira* in the removal plots: (1) 2/9 comparisons of body length, estimated with spiders in the webs ($p < 0.05$); (2) rate of prey consumption, defined as the mean of three determinations made during the season ($p < 0.05$); (3) carapace width of mature spiders at the end of the experiment ($p = 0.05$); and (4) fecundity at end of experiment ($p < 0.05$). Reducing numbers of *Cyclosa* did not affect the density of *Metepeira*, nor its height in the vegetation. Comparable effects on body length were seen for *Cyclosa* in plots from which *Metepeira* had been removed; four out of ten comparisons were statistically significant ($p < 0.05$ or < 0.01). Removing

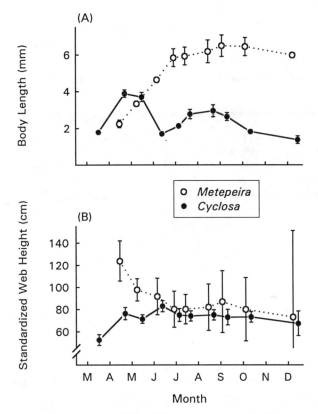

Fig. 4.6 (A) Seasonal change in body size of *Cyclosa turbinata* and *Metepeira grinnelli*. *Cyclosa* completes two generations a year. Overwintering juveniles mature and reproduce in May and June; the offspring of these spiders mature that season, reproducing in late summer or autumn. *Metepeira* is an annual species that reproduces in late summer and early autumn. (B) Change in web height of *Cyclosa* and *Metepeira* over the season. Error bars are 95% CL. (After Spiller 1984a.)

Metepeira had no impact upon *Cyclosa*'s prey consumption rate, final carapace width or fecundity. Thus *Metepeira* suffered more than *Cyclosa* from exploitative competition.

Spiller concludes that *Cyclosa* was more affected by interference competition because most of the statistically significant effects of reducing *Metepeira* numbers relate to *Cyclosa* web placement and occupancy. Removing *Metepeira* produced statistically significant increases in these *Cyclosa* parameters: (1) 1/10 comparisons of web height ($p < 0.05$); (2) 2/10 comparisons of juvenile densities ($p < 0.01$ or 0.001);

Table 4.2. *Spiller's reanalysis of monthly census data from his first-year's experiment (Spiller 1984a) by repeated-measures ANOVA with blocking. The blocks correspond to initial differences between plots in spider densities (described in Spiller 1984a). Treatment and error degrees of freedom are 1 and 2, respectively*

Variable	Treatment	F	p
Cyclosa body length	*Metepeira* removed	6.34	0.064
Cyclosa web height	*Metepeira* removed	554.55	0.0009
Juvenile *Cyclosa* density	*Metepeira* removed	6.68	0.061
Adult *Cyclosa* density	*Metepeira* removed	13.63	0.033
Total *Cyclosa* density	*Metepeira* removed	12.53	0.038
Metepeira body length[a]	*Cyclosa* removed	1.40	0.179

Note:
[a] Possible effects of *Cyclosa* on *Metepeira* densities or web heights were not reanalyzed because none of the individual *t*-tests was statistically significant and no trend appeared in the data.
Source: Spiller, pers. comm.

(3) 3/11 comparisons of adult densities ($p < 0.05$ or 0.01); (4) 2/8 comparisons of number of egg sacs. The absence of effects on prey consumption rate, carapace width and fecundity strongly suggests, as Spiller argues, that web invasions by *Metepeira*, and not exploitative competition for prey, caused the changes in *Cyclosa*'s web height, density and number of eggs sacs, the last-mentioned being related to adult density and the length of time a spider remains at a particular web site. Evidence of web invasions in the food–supplementation experiment support this interpretation (Spiller 1984a).

A potential difficulty in judging the strength of some of Spiller's conclusions is his decision to perform repeated *t*-tests on the same populations at intervals throughout the experiment. Thus tests are not independent; in addition, the actual statistical significance level is lower than stated because of the expectation of obtaining some significant *t*-values by chance when performing numerous tests (Rice 1989). Given these potentials for bias, the relatively small number of significant *t*-tests casts doubt on the strength of the evidence for interspecific competition. Recognizing this problem, Spiller has used repeated-measures ANOVA to reanalyze those data collected at monthly censuses (pers. comm., Table 4.2). The results of his reanalysis support his original interpretation, though not as many variables show statistically significant effects. *Metepeira* clearly had significant effects on *Cyclosa* web height and density, and the effect on *Cyclosa* body length was close to statistical significance. The possibly significant effect of *Cyclosa* on *Metepeira* body length disappeared in the reanalysis. However, the original analysis, which was done on values at the end of the experiment and thus is not subject to the problems of the multiple-census analysis, showed clear effects of *Cyclosa* on *Metepeira*'s feeding rate and fecundity. Even though some of the *F*-values in the reanalysis are not statistically significant, it would be short-sighted to focus solely on probability levels (no matter what the outcome) and not analyze implications of the experiment in the context of the internal consistency of the results and what else is known about the system.

It is intriguing that only *Metepeira* suffered effects of exploitative competition. Spiller explains the unidirectional nature of exploitative competition by arguing that *Metepeira* densities were lower than those of *Cyclosa*. In addition, he offers another quite reasonable interpretation of the asymmetrical exploitative competition: removing *Metepeira* caused the numbers of *Cyclosa* to increase because of reduced interference competition, which resulted in increased intraspecific exploitative competition for prey among *Cyclosa*, which in turn offset effects of reduced interspecific exploitative competition with *Metepeira*.

Cyclosa had the most pronounced impact on *Metepeira*'s feeding rate in March. Spiller explains that only in March are individuals of both species the same size, which correlates with a higher prey overlap in that month (Schoener's 1968 overlap index = 0.88 in March, 0.35 in June and 0.51 in August). In March the species are also vertically separated, which may have contributed to the fact that exploitative competition for prey is unidirectional. Perhaps in early spring most of the prey utilized by both species are small insects that breed in the soil and leaf litter. Since *Cyclosa* are numerous in March and build their webs near the ground, with *Metepeira* higher in the vegetation, the band of *Cyclosa* webs may have intercepted prey that otherwise would have flown upwards into the clutches of waiting juvenile *Metepeira*.

Effects of interference competition also were unidirectional, but opposite to those of exploitative competition, i.e. only *Cyclosa* suffered. Again, the one-way nature of the interaction reflects the consequences of the species' different phenologies. *Cyclosa* densities were higher in the *Metepeira*-removal plots on only three of the 10 dates after removals were started; however, these three dates were during the last six weeks of the season, when *Metepeira* had matured. The size differential between the two species was greatest then, and mature males had started to wander. This is the time of the season when one would predict that *Metepeira* is most likely to invade *Cyclosa* webs. The asymmetry of interference competition in this system follows from the fact that *Metepeira* is much larger than *Cyclosa* during most of the season (Fig. 4.6).

Spiller's study provides convincing evidence of competition between *Metepeira* and *Cyclosa*, but was not designed to measure the intensity of intraspecific competition. Information on competition within each species population would aid in interpreting the results of the removal experiments. Spiller (1984b) conducted another field experiment with his system the next year (1983) in which he tested for both intra- and interspecific effects. He also hoped to obtain more information on interaction mechanisms.

In 1983 Spiller selected slightly smaller plots (1 × 3 m) and expanded the design to a 2 × 2 factorial, in which the two factors were density (unaltered and reduced but not to zero) of either *Metepeira* or *Cyclosa*. This design differed from that of the previous year in that from one-fourth to one-third of the removed species was left in the density-reduction plots in order to test for intraspecific competition. In 1983 Spiller removed spiders daily instead of twice weekly, and instead of one long experiment, he conducted two shorter ones (6–13 April and 6 July–3 August). Spiller calculated an average value over the entire experiment for each parameter and then analyzed the pattern with two-way ANOVA. The evidence for interspecific competition was clear, particularly for exploitative competition affecting feeding rate (Spiller 1984b).

In unaltered plots in the spring, densities of *Metepeira* and *Cyclosa* were *c*. 10 and 7 spiders/m^3 and body lengths were 1.0 mm and 2.5 mm, respectively. Despite these differences in size, the two species consumed prey of similar lengths. Reducing the density of *Cyclosa* significantly increased the number and average size of prey consumed per day by *Metepeira*, resulting in almost a tripling of the biomass of prey consumed daily by the smaller *Metepeira*. Decreasing *Metepeira* densities had no impact on any aspect of *Cyclosa*'s feeding rate. These results agree with those obtained the previous spring, and also are consistent with the finding that intraspecific competition affected the feeding rates of both species.

In summer the spiders build webs at similar heights and *Metepeira* is larger (7 mm body length versus *c*. 3 mm for *Cyclosa*); *Metepeira* now captures significantly larger prey than *Cyclosa*. In the summer of 1983 densities of *Cyclosa* were higher than *Metepeira*'s (9 spiders/m³ versus 4/m³, respectively). *Cyclosa* had no effect on *Metepeira* at this time of the season, but reducing *Metepeira* numbers increased the number of prey consumed per day by *Cyclosa*. The effect was not as strong as the reciprocal influence in the spring. *Metepeira* did not influence the size of prey captured by *Cyclosa*, with the result that reducing *Metepeira* densities did not impact the daily rate of consumption of prey biomass by *Cyclosa*. In contrast to the pattern in the spring experiment, only *Cyclosa* exhibited evidence of intraspecific competition.

These field experiments yielded clear evidence of interspecific competition, yet Spiller concludes that '. . . the ecological significance of the observed processes was ambiguous.' Why? Despite the clear pattern of statistically significant differences in feeding rate, effects of exploitative competition for prey did not translate into detectable changes in density, body length, eggs per sac or total fecundity of the other species. The experiments uncovered no evidence of significant interference competition between *Metepeira* and *Cyclosa*. Some web invasions were documented, but they caused no significant effects on density of heterospecifics in either spring or summer experiment.

Why was evidence for substantial interspecific competition weaker in 1983 than in the previous year? One reason is the shorter duration of the 1983 experiments: one week and four weeks, in contrast to 1982's six-month experiment. Because effects on parameters such as size at maturity and fecundity are cumulative, they will be most apparent in long-term studies. Spiller also suggests that competition was weaker in 1983 because food supply was less limiting that year. He could not test this directly because he did not supplement prey supply in 1983, but comparisons of feeding rates and fecundities would give an indirect indication. *Cyclosa* feeding rates and fecundities did not differ significantly between years, but *Metepeira* exhibited higher rates in 1983 (Table 4.3). Differences in feeding rates are only statistically significant with a one-tailed test, which is not justified because there is no reason independent of the experimental results for hypothesizing a higher feeding rate in 1983. However, the yearly difference in fecundities is clear. The difference between the two species in the degree of food limitation in 1983 agrees with the previous year's food supplementation experiment, in which *Metepeira* displayed the greater response.

Spiller's two years of experimentation uncovered clear evidence of both exploitative and interference competition between two orb

Table 4.3. *Differences between years in prey-consumption rates and fecundity of the orb weavers* Metepeira grinnelli *and* Cyclosa turbinata. *Prey consumption was determined 15–18 August 1982 and 6 July–3 August 1983. Egg sacs were collected 10 September 1982 and a month earlier (4 August) in 1983. All values are from plots in which spider densities had not been experimentally altered*

	Biomass (mg) consumed/ day/spider		No. of eggs/sac/female	
	n	Mean ± s.e.	*n*	Mean ± s.e.
Metepeira				
1982	12	0.39 ± 0.09	10	49.5 ± 3.1
1983	17	0.58 ± 0.07	11	65.6 ± 2.7
t		1.814		3.922
p^a		0.041		0.001
Cyclosa				
1982	23	0.12 ± 0.02	65	19.4 ± 0.7
1983	42	0.18 ± 0.04	21	20.0 ± 1.3
t		0.689		0.480
p		0.247		0.316

Note:
[a] All probabilities are for one-tailed tests.
Source: Spiller 1984b.

weavers. Variation between years in the types of effects uncovered may result partly from differences in experimental design, but also likely reflect variation in prey abundance and other unmeasured factors that influence the intensity of competition. It would be fascinating to subject this system to a removal experiment that continued over several generations.

Additional threads

Often logistical constraints force ecologists to restrict experimental studies to species pairs, even though communities clearly consist of numerous potentially interacting species. One competition experiment with spiders has assessed interactions within an assemblage that more closely resembles a community than the species pairs of the studies that have been discussed so far. Riechert & Cady (1983) conducted a removal experiment with a community of four web-building spiders, which accounted for 98% of the spiders occupying rocky outcrops of sandstone in their Tennessee (USA) study area. These four spiders spin quite different traps. The cribellate lampshade spider, *Hypochilus thorelli*, spins a web of sticky bands of silk. The theridiid *Achaearanea tepidariorum* spins a more irregular tangle web. *Coelotes montanus* is a funnel-web spider that runs down its prey on a flat, non-viscid web. The fourth member of the community, *Araneus cavaticus*, builds a large orb web. Although their webs are different, the four species overlap substantially along the niche dimensions of time of activity, web location and type of prey captured. Three genera – *Hypochilus*, *Achaearanea* and *Coelotes* – are so similar they are 'possibly ecological equivalents' (Riechert & Cady 1983). *Araneus* shows less overlap, primarily because it builds its web farther away from the cliff face than do the others, and it has 'almost exclusive use of concavities in the rock face (bubbles) where it places its lair.' Because of the high degree of niche overlap, Riechert & Cady expected to discover extensive interspecific competition in this community. However, their removal experiment produced no good evidence of competition.

Five different cliff faces were selected: one control (38.4 m long and 2 m high), and four removal areas (dimensions not given; they appear to be similar to the control area). Individuals of three of the four species were removed from the experimental areas, but a different combination was removed from each one. Thus in each removal area a different species was left undisturbed, creating an experimental design that tested for competitive effects from all of the other major web builders in the community on each of the four species. Criteria used for selecting experimental areas were 'close proximity to the control cliff, the same southeast aspect as the control cliff, and the presence of large numbers of individuals of the experimental species.' Starting in August 1980 targeted species were removed weekly until November, and again from March through July 1981. A year of removals produced no statistically significant differences between experimental plots and the control plot, for any species, in population densities (measured as percentage saturation of available habitat on the cliff), egg production or pattern of microhabitat use.

Unfortunately, three problems with the experimental design substantially limit inferences about the presence or absence of interspecific competition in this community. First, apparently high immigration into the experimental plots makes one cautious in concluding that the removal experiment adequately tested for interspecific competition.

Compared to the control area, total densities of heterospecifics were reduced by nearly 50% in the *Hypochilus* and *Araneus* plots, but were not changed appreciably in the *Achaearanea* plot, and were surprisingly 40% higher in the *Coelotes* plot. However, undetected differences in initial densities between control and removal areas could inflate these estimates of immigration. The second problem of interpretation, like that resulting from high immigration, is one encountered in many perturbation experiments. The general problem is that of separating all the consequences of a simple perturbation. In Riechert & Cady's study the difficulty is a consequence of experimenting with species of generalist predators that also prey upon each other, an interaction termed 'intraguild predation (IGP)' by Polis, Myers & Holt (1989). Because almost half of the diet of *Hypochilus* was estimated to be other spiders, removing other spider species may have negatively affected *Hypochilus* by lowering its food supply, possibly counteracting benefits of release from competition. In Chapter 8 we will return to the challenge of separating effects of interspecific competition and intraguild predation.

The third problem with the design of the study is the most serious: experimental treatments and the control plot were not replicated. Riechert & Cady used standard statistical tests to compare parameter values in the removal plots with those of the control. However, using statistics in this manner is an example of pseudoreplication (Hurlbert 1984). The statistical test indicates whether or not the response variable (i.e. fecundity) differs between two plots, but cannot establish whether or not the difference between the control plot and the experimental plot results from the experimental *treatment*. It is likely that logistical constraints made it impractical to replicate plots of the size required for this experiment. High immigration rates suggest that using smaller areas, in which the diluting effects of immigrants would have been even more of a problem, would not have been advisable. This dilemma, whether or not to sacrifice replication in order to obtain less arbitrary units of habitat for experimentation, has continually plagued experimental ecology.

In the absence of replication it is particularly important to know whether initial conditions differ between control and experimental plots. This knowledge does not prevent subsequent 'nondemonic intrusion' of undetected variation in a non-replicated design (Hurlbert 1984), but at least gives some indication of whether or not the sites differ substantially. Unfortunately, Riechert & Cady removed spiders for a month before they censused the plots, and fecundity in the removal plots was not determined at the beginning of the study. Hence it is not possible to compare parameters at the end of the experiment with initial conditions.

Riechert & Cady's (1983) field experiment provides no evidence of interspecific competition, but their results provide no firm evidence against competition either, primarily because of an unreplicated design and the possibility that some of the density reductions were not as effective as desired.

Interspecific aggression and impacts upon population dynamics

Assessing the ecological significance of competition requires placing it in the context of population dynamics. Thus Spiller termed the results of his 1983 experiment 'ambiguous' with respect to their overall significance because of the lack of effects on population parameters. Several ecologists have uncovered interspecific aggressive interactions that are clear examples of interference competition, yet whose population–level consequences are either negligible or difficult to evaluate without additional data.

Hoffmaster (1985b, 1986) sought to uncover possible relationships between aggressive behavior and the structure of tropical orb–weaver communities. She first established that spiders will invade the webs of other species, by releasing 14 *Argiope argentata* and 14 *Nephila clavipes* in the vegetation near an occupied web (Hoffmaster 1986). Of 28 spiders released, 19 invaded the nearby web; seven invaders attempted to evict the resident, and five were successful. It is unclear from this type of study whether or not spiders frequently evict heterospecifics when changing web sites. In another experiment Hoffmaster introduced *A. argentata* or *N. clavipes* to the periphery of a web occupied by either a conspecific or another species, and then determined an index of aggressiveness for the occupant. She discovered that the presence of prey in the web and the degree of starvation of the intruder affected duration of the aggressive encounter. Hoffmaster (1985b) then searched for correlations between the spatial distribution of eight of the species and their relative aggressiveness. She concluded that '. . . overall spider densities could not be predicted from indices of spider aggression obtained by introduction experiments.'

Toft (1987) used a unique experimental approach to examine niche separation between two sheet–web weavers, *Linyphia triangularis* and *L. tenuipalpis*, that co-occur in *Calluna* heath. Despite differences in body size and web dimensions, these two species appear to utilize similar web sites. Rather than measure a multitude of additional variables, which eventually would have yielded differences of questionable ecological significance, Toft demonstrated experimentally that the web sites were essentially identical by transplanting marked adult females to new webs. Transplanted spiders accepted webs of the other species as readily as webs taken from females of their own kind, based upon comparisons of residence times over four weeks after the introductions. During the study Toft observed 49 web take-overs, 22 of which were interspecific; he concludes that '. . . high density situations must lead to competitive interactions of one sort or another.' Additional field manipulations confirm that interspecific web take-overs are not uncommon, with the larger *L. triangularis* displacing the smaller *L. tenuipalpis* (Toft 1988, 1990). The asymmetry of this interference competition leads Toft to suggest that the interaction is a 'producer–scrounger relationship' (*sensu* Barnard 1984), in which individual *L. triangularis* may actually benefit from the presence of the similar, but smaller, congener. These spiders clearly invade and appropriate each other's webs. An important lingering question is the extent to which the presence of each species in a locality affects the other's population density. A removal experiment would be the most direct way to address this question.

Toft has obtained experimental evidence for interference competition between *L. triangularis* and species other than *L. tenuipalpis* (Toft 1986, pers. comm.). The basic experimental unit was a wooden frame, $0.9 \times 0.9 \times 1.8$ m, that was criss-crossed by hemp strings to provide support for web construction. He placed 20 frames on the floor of a Danish beech forest. Natural vegetation grew up within the frames, which were colonized by spiders characteristic of the herb stratum. Toft left 10 units as controls, and from the other 10 he removed all *L. triangularis* that had colonized. Another large sheet-web spider, *L. emphana*, was also removed, but it was unusually rare in the year of the experiment, so its numbers were insignificant compared to the ubiquitous *L. triangularis*. The original goal of the manipulation was to measure competitive release by the smaller, less numerous linyphiids in the herb stratum. Part-way through the experiment Toft observed that juvenile orb weavers were more abundant on the frames from which *L. triangularis* was being removed. He then started to remove juvenile orb weavers from all the units. As before, no linyphiids were disturbed on the control frames.

Removing *L. triangularis* in the spring, when the species was imma-

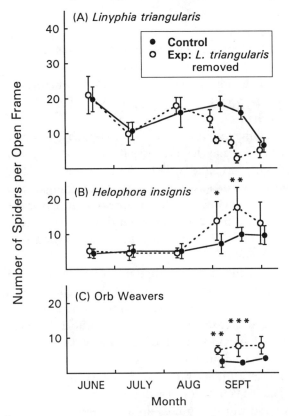

Fig. 4.7 Seasonal variation in number of spiders on experimental units (open frames) from which *Linyphia triangularis* was periodically removed, compared with numbers on control units. (A) Number of *Linyphia triangularis*. All *L. triangularis* were removed on each day that the experimental frames were censused. (B) Number of the linyphiid *Helophora insignis*. *Linyphia hortensis*, which was one-tenth as abundant, appeared on the frames in August and also displayed significantly elevated numbers on the removal frames in September. (C) Total number of orb weavers. Values are means ± 95% CL. One-tailed significance levels (Mann–Whitney U-test) for species other than *L. triangularis*: $\star p < 0.05$; $\star\star p < 0.01$; $\star\star\star p < 0.005$. (After Toft 1986 and unpubl. data of Søren Toft.)

ture, did not significantly reduce its density because young spiders move extensively. Immigration of mature *L. triangularis* in August was also substantial, but Toft was able to maintain their densities on the removal frames below control levels. Numbers of two other linyphiid species were significantly higher on these frames on the two sampling dates in September (Fig. 4.7). Differences were not significant in August, when numbers of L. *triangularis* on the controls had declined so much that they equalled the removal densities. In September total numbers of orb weavers (juvenile and adult *Meta mengei* and juvenile *Cyclosa conica*) were also significantly higher on the L. *triangularis*-removal frames.

Because the spiders were not marked, it is unclear whether L. *triangularis* excluded other species through aggressive encounters or whether the primary mechanism was pre-emption of web sites. Based upon other evidence for interspecific aggressive encounters, it seems reasonable that web invasions may have contributed to the observed pattern, particularly with the linyphiids. It is harder to imagine an adult L. *triangularis* invading the web of a juvenile orb weaver, leading one to speculate that pre-emption of web sites is the major explanation for the competitive interaction between L. *triangularis* and the smaller spiders, including perhaps even the smaller linyphiids. Interspecific predation may also be a factor (Toft 1988).

Toft's studies clearly show that in some situations, particularly where vertical relief of the vegetation may be minimal, sheet-web densities may be high enough for significant interference competition to occur.

Janetos (1982b) concludes that sheet-web weavers and orb weavers were not competing for web sites in a scrubby, second-growth habitat in eastern North America. Abandoned web sites were recolonized by spiders in the same guild at a higher rate than predicted if sites were colonized at random. Janetos concludes that the web sites of the other guild are not attractive and thus the guilds do not compete for sites. Complete identity of sites, i.e. complete niche overlap, is not a necessary condition for competition, and thus his conclusion does not hold. In fact, in both years of his study some abandoned web sites were next utilized by a spider from the other guild, in two cases at a frequency equal to 50% of expected. However, even if both guilds had not distinguished each other's web sites, to conclude that they were competing would have been premature without a field experiment to determine whether or not the supply of web sites is limiting, and whether spiders invaded each other's occupied webs in pursuit of new sites.

Wandering spiders

Only one field experiment has been conducted to test for competition between species of spiders that do not depend upon webs to capture prey. After conducting a long-term study of the spatial and temporal niche relations of the lycosids inhabiting a coastal landscape (Schaefer 1972),

Schaefer (1974, 1975) conducted field experiments to test directly for competitive interactions in one of the salt-marsh communities. Spatial distributions of three of the most abundant species partially overlapped, leading Schaefer to hypothesize that interspecific competition was 'at least partly' the cause of the observed distribution pattern of these species.

Schaefer completed two field manipulations. In one experiment he killed most of the lycosids in four 10×10 m plots by applying parathion in early June; a fifth untreated plot served as a control. Two congeneric species, *Pardosa purbeckensis* and *P. pullata*, rapidly recolonized the treated plots. A larger wolf spider, *Pirata piraticus*, immigrated into the plots much more slowly. By treating the edges of the plots weekly with parathion, Schaefer maintained *P. piraticus* considerably below its density in the control. Densities of all categories of prey in the treated plots had returned to near-normal levels within two months. On 1 October the mean density of *Pirata* in the four treated plots was $9/m^2$ compared with $56/m^2$ on the single control plot. Densities of the two *Pardosa* species did not differ significantly: $38/m^2$ versus $33/m^2$ for *P. purbeckensis*, and $14/m^2$ versus $11/m^2$ for *P. pullata* (means of treated plots versus control density, respectively). Although elevated, averages of the treated plots were not statistically different from the single control values. Statistically significant differences would have been difficult to interpret because the control treatment was not replicated.

In a simultaneously conducted experiment Schaefer treated six 2×2 m fenced plots with parathion and then reintroduced the three species at normal densities, but in combinations to test for competition between *Pirata* and the *Pardosa* species. All treatments were replicated ($n = 2$). Mean densities of *P. purbeckensis* and *P. pullata* were 51% and 15% higher, respectively, in plots without *Pirata*. Schaefer concluded that interspecific competition was not demonstrated, because these elevated densities were not statistically different from the control plots. This conclusion may be too conservative, though, because it is based upon a two-tailed test. A one-tailed test is appropriate, given that the design of the experiment was based upon the hypothesis of negative interactions between the species, which was suggested by Schaefer's extensive previous research on the system. *P. purbeckensis* densities are significantly higher in the plots without *Pirata* under a one-tailed criterion ($t = 3.68$, $p = 0.033$, d.f. $= 2$). The difference for *P. pullata* between removal and control plots is almost statistically significant at the standard rejection level ($p = 0.072$).

Intraguild predation is the likely cause of the negative interaction uncovered by Schaefer's experiment. Based upon data collected on the diet of *Piratus* in non-experimental areas, Schaefer calculated that *Pirata* eats approximately 25% of the progeny of both *Pardosa* species. Schaefer clearly points to predation as a primary component of any negative interaction that his experiments were designed to detect. Predation between species on the same trophic level is often classified as interspecific competition, because the signs of the between-species interaction coefficients are both negative (Abrams 1987b). Schoener (1983a) labels this type of interaction 'encounter' competition. Clarification of commonly used terms is important but debating over what really qualifies as 'competition' could quickly become fruitless. Abrams (1987b) points out that using the term 'competition' to describe interactions that are based on different mechanisms but that have similar population-level consequences has sometimes contributed to confusion in debates over competition. No matter what terminology one elects to use, predation between species on the same trophic level clearly is a different process to exploitative competition for prey, which involves indirect effects that result from consuming shared resources. Evidence for exploitative competition for prey between the lycosid species studied by Schaefer would had to have come from data on individual growth and fecundity, which his experiments were not designed to uncover. Thus Schaefer's results confirm his prediction, based upon spatial overlap and feeding behavior, that *Pirata* has a negative impact on population densities of *Pardosa purbeckensis* and *P. pullata*, but untangling the contributions of intraguild predation and exploitative competition for prey requires further study.

Niche partitioning

Evidence from experimental studies suggests that exploitative competition for prey is not common between species of web-building spiders. Field experiments have uncovered this type of competition in only one system (Spiller 1984a,b). How do spiders, which are food-limited generalist predators, avoid competition? Bristowe (1941) concluded that web-spinning spiders coexist because differences in web structure enable them to capture different types of prey. His explanation rings with the echo of an outdated conventional wisdom that sounds hollow, or at least naive, to modern ears. Nevertheless, Bristowe's postulate can be hammered into more contemporary tones, and tested, by postulating that the observed dietary overlap between coexisting spiders agrees with the quantitative predictions of the now-classic theory of limiting similarity of coexisting competitors (Hutchinson 1959, MacArthur & Levins 1967).

Detailed analyses of spider diets do not provide strong support for a more modern version of Bristowe's conjecture on the mechanism permitting coexistence of web spinners. Dietary overlap between spiders often is greater than what original niche theory predicts should characterize coexisting competitors. High similarity in diet has been found for pairs of syntopic orb weavers, e.g. *Araneus quadratus* and *A. cornutus* (Kajak 1965), and *Araneus quadratus* and *Argiope bruennichi* (Nyffeler & Benz 1978). Nentwig (1985b) discovered that Schoener's (1968) index of niche overlap, defined by prey spectrum, varied between 0.8 and 0.9 for four species of tropical orb weavers.

High overlap in diet is not necessarily inconsistent with simple theories of competitive coexistence; for example, it is possible that prey were not limiting for the population studied by Nyffeler & Benz (1978) and Nentwig (1985b). Species diets are predicted to diverge only if the shared resource becomes limiting. The most direct test of the adequacy of niche separation as a contemporary mechanism of competitive coexistence would be to examine patterns of niche overlap for species that have been studied experimentally. In these systems direct experimental evidence for food limitation and the presence or absence of interspecific competition can be related directly to patterns of niche overlap. Such comparisons reveal the inability of simple niche theory to explain how spiders coexist.

Niche theory and field experiments with spiders

The basilica and labyrinth spiders highlight the failure of classical niche theory to explain competitive coexistence in web-building spiders. These araneids build webs as different as any pair of syntopically distributed orb weavers. The labyrinth spider constructs a vertically oriented viscid catching spiral, with openings of *c.* 2 × 5 mm. In contrast, the basilica spider's orb is a non-viscid dome, with openings approximately one-tenth as large as *Metepeira's*. Field experimentation uncovered no evidence of exploitative competition for prey between these species (Wise 1981; this chapter). It is reasonable to hypothesize that these spiders avoid competition because they hunt with such different traps. Surprisingly, though, their diets are remarkably similar (Wise & Barata 1983;

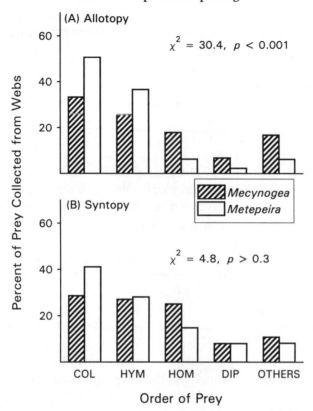

Fig. 4.8 Overlap in diets between the labyrinth spider (*Metepeira*) and the basilica spider (*Mecynogea*). Numbers of *Metepeira* prey identified were 162 and 106 for the allotopic and syntopic populations, respectively; corresponding sample sizes for *Mecynogea* were 149 and 90. COL = Coleoptera; HYM = Hymenoptera; HOM = Homoptera; DIP = Diptera; OTHERS = Odonata, Lepidoptera, Hemiptera, Orthoptera and Neuroptera. The apparently greater similarity of diet in syntopy than in allotopy is not statistically significant [$p(\chi^2) > 0.2$ for the interaction term in a $2 \times 5 \times 2$ contingency table]. Interestingly, divergence between allotopic and syntopic diets appears to be greater at the family level, particularly for the two most abundant families of beetles in the diet. (After Wise & Barata 1983.)

Fig. 4.8). Syntopic populations did not differ significantly in the taxonomic composition of the diet. Pianka's (1973) index of niche overlap for the syntopic populations, based upon classification of prey to order, was 0.96, which is too high to permit coexistence of two competitors. Sample sizes were not large enough to calculate the overlap index based upon number of prey per family; however, broad similarity of diet in syntopy is also seen when prey are identified to the family level (Wise & Barata 1983).

Partitioning of prey does not explain why spiders in the rock-face community investigated by Riechert & Cady (1983) were not competing. The four species that they manipulated were in four different families, yet there was considerable dietary overlap; pair-wise coefficients of similarity, based upon taxonomic category and size of prey, ranged from 0.45 to 0.81. Overlap in this community is considerably less than that between the basilica and labyrinth spiders, and is in the range of the criterion suggested by MacArthur & Levins (1967) for competitive coexistence. Unfortunately, it is not possible to relate these niche differences to the absence of competition because the design of Riechert & Cady's removal experiment prevents strong firm conclusions about the absence of competition. However, in agreement with our interpretation of the major cause for differences in diet between basilica and labyrinth spiders, Riechert believes that most differences in diet in the rock-face community reflect 'location on the cliff face rather than web type itself' (Riechert & Łuczak 1982).

It would be misleading to imply that differences in web structure are entirely unrelated to differences in the size and taxonomic composition of prey captured. For example, Pasquet (1984b) concluded that divergence in diet of the vertically segregated *Argiope bruennichi* and *Araneus marmoreus* reflects differences in web design and spider behavior more than differences in web location, since prey availability was similar in the two different vegetation strata utilized by the species. Eberhard (1990) gives other examples. Most evidence, however, suggests that differences in web structure, particularly among closely related species, do not lead to substantial divergence in diet. After reviewing the literature, Eberhard (1990) concluded that '. . . differences in details of orb designs probably do not represent fine tuning to specific subsets of prey.'

Studies of variation in dietary overlap between the two common North American species of *Argiope* provide some of the best evidence against invoking divergence in web structure as a mechanism promoting competitive coexistence. Uetz *et al.* (1978) related statistically significant differences in taxa and size of prey captured by the two similar old-field orb weavers *Argiope aurantia* and *A. trifasciata* to the former's larger body size and the more open mesh of its web. In contrast, however, Taub (1977) found the diets of these two species not to differ significantly. Also, Brown (1981) uncovered no consistent pattern when he compared dietary overlap between these species in several different areas. McReynolds & Polis (1987) also could find no obvious patterns in prey capture; mesh density had no clearly discernible effect on the type of prey captured – certainly not enough to indicate significant niche partitioning.

Causes of the inconsistent pattern become apparent when results of field experiments with these two *Argiope* spp. are interpreted in light of niche shifts that occurred during the second year of our research (Horton & Wise 1983). Interspecific competition was absent both years. Initially dietary differences between *A. aurantia* and *A. trifasciata* appeared tantalizingly consistent with the predictions of classical competition theory, but an additional year of field experimentation forced us to revise our interpretation.

In our study area the species differed in the taxonomic composition of prey captured. *A. trifasciata* tended to capture small prey (Homoptera), whereas *A. aurantia* had a greater proportion of the larger Orthoptera and Hymenoptera in its diet. Pianka's (1973) overlap index, calculated for the number of individuals in each prey taxon, was 0.69 during the first year. This overlap is considerable, but is not inconsistent with the interpretation that these two potential competitors are avoiding competition for prey through partitioning of their food resources. Thus, the absence of competition was compatible with an interpretation that postulated its importance over evolutionary time. This simple, single-factor explanation of the reason for dietary differences disappeared when we repeated the field experiment the next year. Niche overlap was higher in the second year of the study, despite changes in the environment that suggested competitive pressure should have intensified.

As originally derived, classical niche theory predicts that niches should diverge in response to increased pressure from interspecific competition. Since decreased availability of prey should intensify exploitative competition, two similar species should exhibit decreased niche overlap in response to decreased prey availability. Schoener (1983a) lists 30 cases in which overlap in resource use decreased during times when resources were less available. We could test for such a response because prey were apparently less abundant the second year of the experiment. We did not measure prey availability directly, but several facts suggest that prey supply was more limited in 1980. First, the summer of 1980 experienced a drought relative to 1979; average rainfall for July through August was 18.7 cm and 38.8 cm, respectively. The previous 10-year average for these two months was 33.0 cm. It is reasonable to hypothesize that the drought lowered insect productivity. Secondly, in 1980 *A. aurantia*'s rate

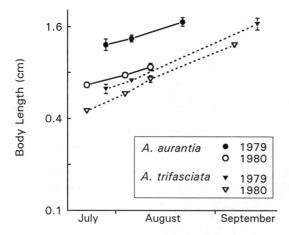

Fig. 4.9   Changes in mean body length (± s.e.) of *Argiope aurantia* and *A. trifasciata* over the season for both years of the experiments; density and species-removal treatments have been pooled. (After Horton & Wise 1983.)

of prey capture, weighted by prey length, was 50% of its 1979 value. *A. trifasciata*'s rate was unchanged. Finally, growth rates of both species were lower in 1980 (Fig. 4.9). This effect was particularly pronounced for *A. aurantia*, in agreement with its lower estimated rate of prey capture.

Argiope aurantia and *A. trifasciata* behaved contrary to the predictions of classical competition theory. When food supply decreased, their diets converged (Fig. 4.10). Pianka's (1973) overlap index calculated for prey taxa changed from 0.69 in 1979 to 0.98 in the year of the drought. Convergence is also apparent when indices are weighted by the mean length of prey in each taxon: 0.76 and 0.90 for 1979 and 1980, respectively. A major cause of the convergence in diet is the increased proportion of Homoptera in the diet of *A. aurantia*, and the dramatically decreased fraction represented by Orthoptera. Convergence in diet is correlated with greater similarity in web-site characteristics in 1980 than in the previous year. In both years the species showed considerable overlap in web height and the size of opening in the vegetation utilized, with *A. trifasciata* tending to locate its web higher in the vegetation and in smaller openings. In 1980, possibly in response to increased risk of desiccation, both species placed their webs lower in the vegetation and spun them in smaller openings. These shifts were more pronounced for *A. aurantia*, which is less resistant to desiccation than its congener (Markezich 1987). Both species altered their foraging, with consequent

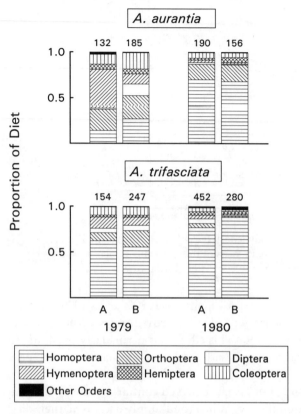

Fig. 4.10 Major insect orders in the diets of *Argiope aurantia* and *A. trifasciata*. 1980 was the year of the drought. Data for all plots within blocks A and B have been pooled. Numbers above each column are the total number of prey classified to order. (After Horton & Wise 1983.)

effects on growth, more in response to physical factors and vegetation structure than in response to competitive interactions with each other. This response suggests that the absence of competition between these species is not the result of differences in their food niches.

Thus patterns of niche overlap between the two *Argiope* species, and between the basilica and labyrinth spiders, are inconsistent with the simple predictions of niche theory as originally conceived. This lack of correspondence suggests that the classical competitionist paradigm is inadequate to explain niche relationships among web-building spiders. Niche theory itself has become much more complex since the early quantitative formulations of Hutchinson, MacArthur and Levins. For example, it is clear that it is not reasonable to expect that a single value of niche overlap will define the minimum separation required for coexistence for all situations in which interspecific competition is strong (Schoener 1982, Abrams 1983). In fact, constraints in addition to those imposed by interspecific competition (i.e. differences in relative prey profitabilities, intense intraspecific density dependence) could lead to niche convergence between competitors (Schoener 1982, Abrams 1986). No necessary connection even exists between measures of niche overlap and the intensity of competition (e.g. Colwell & Futuyma 1971).

Spiller's experiments, which revealed interspecific competition between *Cyclosa* and *Metepeira*, illustrate well the shortcomings of the over-simplified world view of classical niche theory. Spiller has shown that, for the system he investigated, it is unreasonable to expect that a single value of niche overlap could define the limits of similarity that might permit coexistence. First, the shift in relative size of the two species over the growing season, with consequent shifts in competitive advantage and the mechanism of competition (i.e. exploitative versus interference), invalidates the use of a single index of niche overlap. Secondly, even within a season the simple index of niche overlap proposed by MacArthur & Levins (1967) to estimate the competition coefficients does not adequately predict the outcome of exploitative competition for prey (Spiller 1986a). The MacArthur & Levins index simply reflects the relative extent to which diets overlap. Schoener (1974) proposed a more realistic index of competitive interaction that weights the proportion of each food item in the diet of each species by its relative frequency in the environment, its rate of consumption by each species, and the efficiency with which it is assimilated. Spiller (1986a) calculated both types of overlap index for *Cyclosa* and *Metepeira*, and found that the indices markedly differed. Schoener's formula was a much better qualitative predictor of the direction and magnitude of exploitative competition for prey between *Cyclosa* and *Metepeira* as measured directly by changes in feeding rates that occurred in response to experimental manipulations of competitor density (Spiller 1984b, 1986a).

Non-experimental studies

Chapter 3 explored the competitionist paradigm in the research of spider ecologists, who interpreted patterns in the context of classical competition theory, defined as the early quantitative theories of limiting similarity or the antecedent qualitative versions. This research was non-experimental. Results of field experiments seriously call into question whether conventional competition theory is applicable to spider communities. Some non-experimental studies of spider communities have also uncovered patterns of niche overlap that do not support a simple competitionist interpretation of spider community structure.

Maelfait *et al.* (1980) tested the hypothesis of niche complementarity in four spider communities and found that species which overlap highly in the niche dimensions of either habitat or spider size (i.e. they exhibited a particularly low Hutchinsonian ratio) are not likely to be more separated in the temporal niche dimension. In a study of forest-floor web builders, Leclerc & Blandin (1990) found patterns of species co-occurrence that do not support the hypothesis that interspecific competition is important in organizing this spider community. Hoffmaster (1985a) compared the structure of temperate and tropical forest orb-weaver communities in the context of the niche compression hypothesis, which states that more intense competition in tropical communities has led to the compression of tropical niches over evolutionary time (Dobzhansky 1950, Pianka 1983). Hoffmaster measured the niche dimensions of web height, vegetation height, height of the web supports, percentage canopy cover, and vertical and horizontal distances of the web to the nearest vegetation. These parameters discriminated occupied sites from random points in the forest. Niche breadth was defined in terms of the dispersion of points along each of six axes of a Principal Components Analysis. Hoffmaster found no difference in niche breadths in the temperate (seven species) and tropical (nine species) communities, leading her to conclude that '. . . competition for web sites is not important in structuring (these) orb-web communities.'

Shelly (1984) argues that the vertical distribution of the neotropical orb weaver *Micrathena schreibersi* does not reflect interspecific competition. Prey are most abundant near the ground, yet *M. schreibersi* webs are most common higher in the vegetation. No similarly sized web builders place webs near the ground in *M. schreibersi*'s habitat. Shelly speculates that *M. schreibersi* may build its webs higher in the vegetation not to avoid competition, but to escape ground-dwelling predators.

Schaefer (1972) studied for over six years the spatial and temporal distributions of 11 wolf spiders in 17 different habitat types in a coastal landscape. He concluded that niche separation prevents current interspecific competition from being a factor for most lycosid species pairs, with the possible exception of interactions between *Pardosa purbeckensis* and *P. pullata*. He argues that most negative interactions between lycosids in nature result more from intraguild predation and direct confrontation than from exploitative competition. He concludes that '. . . Konkurrenz kein Hauptfaktor bei der aktuellen Einnischung der Lycosiden in die Landschaften des Bottsandes ist' ('. . . competition is not a major factor in determining ongoing niche relationships between lycosids in the Bottsand Landscape').

Jocqué (1984) discusses a broad geographical pattern in the distribution of soil-dwelling web and wandering spiders in the context of two quite different hypotheses. As an example of a general pattern, Jocqué cites Enders' (1975b) observation that the percentage of the total spider fauna comprised by linyphiids increases from the equator to higher latitudes, whereas the relative contribution of salticids decreases in habitats farther from equatorial areas. Enders (1975b) proposed that this pattern results from exploitative competition, with salticids being more efficient competitors than linyphiids in tropical environments, where small prey such as Collembola are hypothesized to be rarer. Jocqué rejects Ender's hypothesis after comparing spider faunas in different temperate and tropical habitats and analyzing differences in life cycles. Jocqué favors the hypothesis that low temperatures limit the distributions of many jumping spiders to tropical climates, and that linyphiids are limited by interference competition from other groups, such as ants, which are more numerous in tropical than temperate regions.

Competition, resource partitioning and evolutionary time

Could responses to competition over evolutionary time reconcile the competitionist view of spider communities with the results of field experiments, which either fail to uncover current competition or have found competition in a system in which niche relationships do not fit the simple theory of limiting similarity? Research by Turner & Polis (1979) illustrates the difficulties inherent in such an attempt at reconciliation. These investigators reached ambivalent conclusions on the status of interspecific competition after they studied five species of crab and lynx spiders inhabiting inflorescences. They calculated niche overlap along several traditional niche dimensions: season, prey species, prey size and habitat (species of shrub occupied). Average niche overlap was high along all dimensions for one species pair; however, for the remaining nine pair-wise comparisons overall niche overlaps were less than the 0.54 criterion for competitive coexistence proposed by MacArthur & Levins (1967). This observation, along with evidence of niche complementarity for most pairs, led Turner & Polis to conclude that 'it is unlikely that widespread competition occurs among the members of this guild,' presumably because past competition has favored the evolution of niche partitioning. However, they then explain the absence of interspecific competition by a superabundance of resources, arguing that inflorescences were not limiting because 5% were occupied. They also reason that prey was not limiting because spiders are euryphagous, and also because spiders do not limit their prey populations. These latter two arguments are not compelling, particularly in light of evidence from Morse & Fritz's (1982, 1987) research that a shortage of prey often limits thomisid growth and reproduction. Nevertheless, competition could be absent even in the presence of severe prey limitation by the spider guild; i.e. prey shortages need not necessarily operate as a density-dependent factor (discussed in detail in the next chapter). Turner & Polis conclude that inter- and intraspecific predation keeps densities low enough to prevent competition. However, like many non-arachnological colleagues, they also invoke Connell and Rosenzweig's Ghost of Competition Past (Connell 1980, 1985) by proposing that '. . . competitive interactions (during ecological crunches) could at least periodically contribute to the observed patterns of coexistence.' Thus, they hypothesize the importance of exploitative competition despite their belief that prey levels are too high for competition, and despite the postulated importance of predation in keeping spider densities below competitive levels.

Non-experimental data used by some ecologists to argue that competition and resource partitioning are important leads others to doubt the existence of competition, or at least to doubt that it is widespread or shapes niche differences between spiders. Part of the confusion comes from the lingering problem of how to differentiate between evolutionary and ecological time scales. On occasion ecologists appear unclear whether their conclusions relate to ecological time scales, evolutionary time scales, or both. A rigid distinction is unreasonable, but an operational distinction for the purposes of hypothesis building and testing is crucial.

Evolutionary theory predicts that even small selective advantages can lead to substantial adaptive changes in populations, given sufficient time. This prediction gives a theoretical handhold for clinging to classical competition theory even when field experiments fail to uncover interspecific competition over ecological time, or when arguments from non-experimental data lead to the conclusion that competition between species is absent or of minor importance. Connell (1980) presents a demanding set of requirements to be satisfied in order to establish that current niche differences are adaptations to past competitive pressures. The burden of proof proposed by Connell to establish that interspecific competition has been a major cause of observed niche differences between species is awesome, requiring information on ancestral populations and proof of genetic divergence. No example exists for spiders that satisfies all of his criteria. In fact, these requirements have been satisfied for few, if any, pairs of plant or animal species. Niche differences that evolved in response to non-competitive selective pressures could still favor coexistence, but why focus on interspecific competition to explain community patterns if experimental evidence suggests that it is absent or weak over ecological time, and if the criteria for proving its past evolutionary importance are nearly impossible to satisfy for most systems?

Competitive ghosts no longer haunt the hypotheses of spider ecologists, as they did the theories of Tretzel (1955) and others. Several developments, some in the field of community ecology at large, have accomplished this exorcism. The continuing refinement of the use of neutral models in community ecology is supplanting simple appeals to differences in niche overlap as evidence of competitive displacement. In addition, more ecologists are coming to recognize that consequences of competition may differ depending upon the mechanism involved and the time scale of the interaction (e.g. Abrams 1990). With respect to spiders, in the one situation in which clear experimental evidence of competition between spider species exists (Spiller 1984a,b, 1986a), niche differences that promote coexistence are unlikely the result of co-evolution between the two species. It would be difficult to argue that the seasonal reversal of competitive advantage, which results from the markedly different phenologies of *Metepeira* and *Cyclosa*, reflects an intimate co-evolutionary history. Thus no ghost is needed to explain niche patterns for the best example to date of interspecific competition between spiders. Furthermore, niche partitioning is not a consistently convincing explanation for the absence of competition between species of web-building spiders in those situations in which field experiments have uncovered no evidence of current competition.

Field experiments suggest that current interspecific competition between spiders is uncommon, though this conclusion is limited to web spinners. Whether or not wandering spiders compete, particularly through exploitative competition for prey, has not been adequately tested. Nevertheless, the issue of competition in spider communities now centers on why web-spinning spiders usually do not compete. Co-evolved niche differences appear not to be the explanation; we must turn elsewhere for an answer.

Synopsis

The competitionist paradigm has determined the direction and focus of numerous ecological studies of spiders, with many ecologists designing research programs centered on resource limitation and competition. Patterns in niche overlap, foraging behavior and population density have been attributed to the impact of interspecific competition. Several investigators, however, have tempered their interpretations by cautioning that direct experimental evidence of competition between spiders was lacking. Recent manipulative field experiments generally have failed to support competitionist views of communities of web-spinning spiders.

An early test of competition for web sites between two linyphiids failed to find evidence of competition when individuals of one species were added to open experimental plots. In a later study, mixed- and single-species populations of two forest orb weavers were established on open experimental frames. These species, the basilica and labyrinth spiders, have similar phenologies, and a previous field experiment demonstrated that a shortage of prey limited their fecundity; yet the density manipulation uncovered no evidence of interspecific competition. Presence of the other species did not affect survival rates on the experimental units nor rates of egg production for either species. Field experiments with two similar congeneric old-field orb weavers also uncovered no evidence of interspecific competition. Low and high densities of single- and mixed-species treatments were established on natural vegetation in open plots. Two years of experimentation uncovered no evidence that interspecific competition affects web placement, feeding rate, individual growth rate or survival.

Another study examined four species that represented most of the web spinners inhabiting a series of rock faces. The experimenters removed all species but one, in different combinations, and looked for evidence of release from competition in the remaining species, when compared to an undisturbed control cliff. No evidence of interspecific competition emerged from the study, though lack of replication and no knowledge of initial conditions considerably weaken the findings.

In contrast to these findings, field experiments on a levee bordering a salt marsh uncovered clear evidence of competition between two species of orb weavers that differ markedly in phenology and adult body size. A supporting study demonstrated that a shortage of prey limits the fecundities of both species. In the first year of the research, removal experiments on open plots uncovered evidence that exploitative competition for prey affected the feeding rate and fecundity of the larger species, but not the smaller. The larger species appeared to interfere with the smaller by invading its web. Competitive advantage reversed during the season as the relative sizes of the species switched. The second year's manipulation produced clear evidence that exploitative interspecific competition affected feeding rates of both species. However, neither species affected the density, individual growth rate nor fecundity of the other. Thus effects of competition on feeding rate did not translate into effects that determine population density. This may indicate that interspecific competition was not a major interaction that year, or it may reflect the short duration of the second year's experiments.

Interference competition may be important among web spiders in some situations. The presence of a large linyphiid had a negative impact upon the rates of colonization of open experimental frames by a smaller linyphiid species and juveniles of an orb-weaving species, probably through a combination of web invasions and pre-emption of web sites.

Accumulating experimental evidence does not support the hypothesis that interspecific competition for prey is a significant interaction in most communities of web-building spiders. Among wandering spiders the hypothesis of exploitative interspecific competition has not been adequately tested. A set of field experiments in a salt marsh uncovered a negative interaction between lycosids that is most likely intraguild predation and not competition for prey. The experiments were not designed to test for exploitative competition. The question of whether species of wandering spiders compete with each other for prey has not yet been addressed directly with field experiments.

Resource partitioning frequently is proposed to explain the absence of interspecific competition over ecological time in systems in which one expects to find it: species avoid competition because past competitive pressures have caused niches to diverge to the point where current competition is weak. Such an explanation does not satisfactorily explain the absence of interspecific competition in most field experiments with spiders. The basilica and labyrinth spiders have distinctly different catching spirals, the type of difference that often has been advanced as an example of evolutionary divergence that minimizes dietary overlap and permits coexistence. However, these spiders capture remarkably similar prey. The two old-field orb weavers (*Argiope* spp.) that were not competing with each other did differ significantly in the taxonomic composition of prey captured in the first year of the study. In the second year prey abundance appeared to limit growth more than the previous year, yet the diets of both species converged. Classical competition theory predicts the opposite pattern, and thus does not explain the absence of competition in these systems. Even in the one system where current competition for food between spider species has been demonstrated (*Cyclosa* and *Metepeira*), niche relationships are more complex than predicted by simple niche theory.

Interspecific competition for prey appears not to be widespread among web-building spiders. Co-evolved niche separation is not the most reasonable hypothesis to explain the absence of current competition; thus, we must examine other possible explanations for how spiders avoid competition.

5 · How spiders avoid competition

Prey scarcity as a density-independent limiting factor

Accumulating evidence supports the hypothesis that a relative shortage of prey affects growth and fecundity of web-building spiders by acting as a density-independent limiting factor. Thus, though food supply is often limiting, exploitative intraspecific competition is absent or weak. Consequently, exploitative competition for prey between different species of spiders is even less prevalent, given that two species will differ in resource requirements because they are different gene pools. Below I present experimental evidence that interspecific competition among web- building spiders is infrequent because *intra*specific competition for prey is a minor interaction.

Old-field orb weavers

Manipulating *Argiope* densities (Horton & Wise 1983) failed to uncover evidence of substantial intraspecific competition. Some negative intra-specific density effects appeared during the first year of the study (1979). Increasing the density of conspecifics increased the mean web height of both species, but the differences between treatments (33% for *A. trifasciata* and 74% for *A. aurantia*) were statistically significant only for the census immediately following the density manipulations. *A. aurantia*, but not *A. trifasciata*, had a significantly higher growth rate in plots in which intraspecific density was lower (proportional changes in length over the season were 1.7 and 1.3, respectively). Survival, defined as the net effects of mortality and migration, was significantly density dependent for *A. trifasciata* when calculated to the last pre-reproductive census ($p < 0.05$), but was non-significant when calculated to the census 2.5 weeks earlier (F of ANOVA $= 0.44$). I have argued that the change in niche overlap observed during the two years of this study is inconsistent with a competitionist interpretation of niche differences between the two species. Yet in the absence of interspecific competition, evidence that intraspecific competition affects web placement, growth and survival constitutes a strong argument for the importance of niche differences in preventing competition. In principle this argument is compelling, but unfortunately for the competitionist paradigm the evidence for intraspecific competition in *Argiope* spp. is not as strong as it first appears. Several inconsistences weaken the case for substantial intraspecific competition in this system. One incongruity is the absence of evidence of intraspecific competition in the year in which the experiment was repeated, despite the indirect evidence that prey supplies were scarcer this year (Chap. 4). Furthermore, if intraspecific competition was strong the previous year, why were the same parameters not consistently affected in both species? The most parsimonious explanation is that confounding variation contributed to many of the statistically significant treatment effects in 1979, the first year of the study.

In 1979 we observed substantial variation between plots in spider size immediately following the density manipulations (Horton & Wise 1983). In this year resident spiders were left in many of the treatment plots. In addition, the vegetation in one high–density plot was substantially higher than in any of the other plots. Thus apparent intraspecific density effects in 1979 could have resulted from differences in the height of substrate for web construction and initial differences in spider size unrelated to the treatments. In 1980 we removed this possibly confounding variation by modifying our random assignment of treatments to plots, and by clearing all plots of spiders and then introducing spiders to the plots at random (Horton & Wise 1983). Initial web heights and body lengths were less variable between plots in 1980; and in this year no negative intra– or interspecific density effects were uncovered. Some negative density effects in 1979 could have been non–artifactual. This is particularly true for effects on growth, for which initial size differences were taken into account in the ANOVA. Considered as a whole, however, the results of the two years indicate that exploitative intraspecific competition for prey was only a minor interaction in both *Argiope* species.

Forest orb weavers

Absence of experimental evidence for competition between two forest-dwelling orb weavers, *Mecynogea* and *Metepeira*, was also accompanied by evidence of only weak intraspecific competition. I concluded that intraspecific competition was unimportant because density of conspecifics explained only 5% of the variance in web height of *Mecynogea* at the beginning of the experiment, 2% of the variance in the number of days female *Metepeira* remained in the experimental populations, and 1% of the variance in number of eggs per sac of *Mecynogea*. In fact, evidence for intraspecific competition is even weaker than I originally concluded, because in my correlation and multiple regression tests I used an inappropriate error term (Hurlbert 1984). I analyzed values for individual spiders, many of which came from the same high-density experimental populations; instead, I should have examined the correlation between mean densities per experimental unit with mean survival and fecundity rates for each unit. I decided against this approach for analyzing fecundity because the only fecundity measure that was independent of survival time was the mean rate of egg production per day. However, this statistic shares the same denominator, i.e. mean days on the unit, with the average density on the unit (Wise 1981). A correlation between two ratios can be a statistical artifact if the ratios have the same denominator (Atchley, Gaskins & Anderson 1976). Thus in avoiding a potential hazard I jumped from the pot of hot water into the statistical fire, but was only slightly singed because I concluded that the statistically significant effects were of minor importance because they were so small.

Originally I hypothesized that intraspecific exploitative competition for prey was unimportant because prey were not limiting for either species in the year that I tested for competition (Wise 1981). The previous year I established that supplementing prey increased the fecundity of *Metepeira* and *Mecynogea* (Wise 1979), but was unable to test adequately for density effects. In the year of the competition experiment I did not test for prey limitation, but noted that reproductive rates of both species that year resembled those of spiders receiving supplemental prey the previous year. Thus I concluded that competition was absent the year of the competition experiment because food was not a limited resource. Whether or not interspecific competition occurs in years in which a limited prey supply induces intraspecific competition in *Metepeira* and *Mecynogea* populations remained an unanswered question. I designed an experiment for the next year in which I would simultaneously test for prey limitation and competition, both intra- and interspecific. Unfortunately, the intervening winter was the coldest in a century, and mortality of *Mecynogea* spiderlings in egg sacs was > 90% (unpubl. data). Thus the weather gods encouraged me to restrict the investigation to an examination of the relationship between the degree of food limitation and intraspecific competition in *Metepeira*, which, unlike *Mecynogea*, is not near the northern limit of its geographic range in Maryland.

In these experiments I modified the design of the open experimental unit used to support populations of *Metepeira*. In the previous competition experiment I was forced to use a range of spider densities because highly variable rates of colonization of the dead branches on the units made a factorial design difficult to achieve. An alert field assistant, David Weiss, noticed that both basilica and labyrinth spiders lived on abandoned metal fencing along the road to our research areas. Taking inspiration from Arachne, we constructed prototype wood frames that held 2.5-cm mesh poultry fencing as web supports, and discovered that added spiders colonized at a high rate, as did non-introduced species from the surrounding vegetation. The pilot experiment, started just as remnants of Hurricane David visited the study site, achieved a colonization rate of over 90% of introduced spiders. This success encouraged us to use this type of experimental unit in future experiments, enabling us to achieve designs with clear-cut low and high spider densities. Thus we hoped to avoid some of the previously cited statistical difficulties.

We demonstrated that intraspecific competition was absent even though prey was a limited resource for *Metepeira* (Wise 1983). The experiments revealed no negative density effects on survival, growth or fecundity, despite evidence that prey shortages reduced growth and rates of egg production. Supplementing the prey of juvenile spiders increased their growth rate, but growth rates were similar at high and low densities (Table 5.1). Neither prey abundance nor spider density influenced juvenile survival. A separate experiment was conducted later in the season with marked adult females. Survival, number of eggs laid per sac, and net reproduction over the entire reproductive season were independent of *Metepeira* density (Table 5.2). I did not test directly for food limitation of adults, but indirect evidence suggests they, like earlier instars, were also food-limited that year. Mean number of eggs per sac was significantly lower than in the year in which experimentally increasing the food supply increased fecundity (14 ± 1 eggs per sac versus 22 ± 3, respectively). Hence it is clear that intraspecific competition for prey was absent even when prey was limiting for the labyrinth spider.

Filmy dome spider revisited

Absence of intraspecific competition for prey may be widespread among web-building spiders. In my first experimental study of prey limitation

Table 5.1. *Results of field experiment with juvenile* Metepeira labyrinthea *to test simultaneously for food limitation and intraspecific competition*

Effect tested	Treatment[a]	Initial density (mean no./ unit ± s.e.)	Proportion surviving[b] (mean ± s.e.)	Final live weight (mg)[c] (mean ± s.e.)
Food limitation	Natural prey (low-density and high-density units pooled)	18 ± 5^d	0.18 ± 0.03	15 ± 2
	Supplemental prey (high-density + prey units)	69 ± 7	0.19 ± 0.09	24 ± 3
	t-statistics[e]		0.14 n.s. (d.f. = 21)	1.71★ (d.f. = 16)[f]
Density	Low-density	6.3 ± 0.3	0.16 ± 0.03	15 ± 2
	High-density	67 ± 9	0.23 ± 0.02	16 ± 2
	t-statistics		0.29 n.s. (d.f. = 19)	0.36 n.s. (d.f. = 14)

Notes:
[a] The design consisted of three types of open experimental units: low density ($n = 19$), high density ($n = 4$) and high density + supplemental prey ($n = 2$).
[b] Proportion surviving = (no. of females removed on Day 25)/(total no. of spiders, all of which were immature, on Day 0).
[c] Females only. On Day 25, 42 of the 107 females removed from all the units were mature.
[d] Standard error is large because low- and high-density units have been pooled in order to satisfy the statistical requirement of orthogonal comparisons when testing for both food limitation and density effects. No spiders actually experienced a mean density close to 18 spiders per unit at the start of the experiment.
[e] n.s. = not significant; ★ = significant at the 0.05 level (one-tailed criterion).
[f] Females from units receiving supplemental prey tended to be larger as measured by length of the fourth tibia, but the difference was not statistically significant (1.27 ± 0.04 mm versus 1.11 ± 0.05 mm; $p(t) > 0.1$).
Source: Wise 1983.

and competition among spiders, I concluded that a shortage of prey acted as a density-independent limiting factor in its effect upon the growth rate of early instars of the filmy dome spider, *Neriene radiata* (Wise 1975). However, I also concluded that competition for prey among adult females lowered their fecundity and increased the rate at which they emigrated from the open experimental populations that I had established on small ground junipers. This discovery of intraspecific competition, in conjunction with the hypothesized importance of

Table 5.2. *Results of field experiment in which densities of mature female*
Metepeira labyrinthea *were manipulated to test for effects of competition*
upon survival and fecundity

Treatment[a]	Initial density (mean no./unit ± s.e.)[b]	Proportion surviving (mean ± s.e.)[c]	Mean no. eggs/sac ± s.e.	Net reproduction (mean no. eggs/ female ± s.e.)[d]
Low density	2.9 ± 0.2	0.70 ± 0.07	14.3 ± 1.0	30 ± 4
High density	30 ± 2	0.68 ± 0.03	14.2 ± 0.6	35 ± 5
Test statistics		$U = 38$ n.s.[e] ($n = 3,25$)	$t = 0.01$ n.s. (d.f. = 23)	$t = 0.41$ n.s. (d.f. = 26)

Notes:
[a] 25 low-density populations, 3 high-density populations.
[b] No. of individually marked spiders in webs on the frames on 21 August 1980.
[c] Proportion of marked spiders present on the frame on Day 28 of the
experiment (18 September). Survival to the end of the experiment is
incorporated into net reproduction.
[d] Net reproduction calculated to Day 90. On this date all egg sacs were removed
from the frames; mean proportions of original females present on Day 90 were
0.38 ± 0.07 and 0.34 ± 0.02 for low- and high-density units, respectively.
[e] Mann–Whitney U-statistic; t-statistic was not calculated because variances were
not homogeneous after data had been transformed. Median proportions
surviving were 0.75 and 0.65 for low- and high-density populations, respectively.
Source: Wise 1983.

competition in spider communities and the general reliance on the
primacy of resource competition in general theory, gave impetus to my
later experimental studies with spiders of interspecific competition
(Wise 1981, Horton & Wise 1983, Wise & Barata 1983). Failure to find
substantial inter- or intraspecific competition in subsequent experiments
has prompted me to re-evaluate the strength of the evidence for
intraspecific competition between adult female *N. radiata*. The result is
surprising: **no evidence exists for intraspecific competition among
filmy dome spiders**. Apparent negative effects of density on survival
and fecundity in my original analysis were artifacts of using inappropriate models to analyze the data.

In these experiments I located several dozen ground juniper bushes of
approximately 1 m² in area; webs of filmy dome spiders were often quite
dense on these junipers. The sheet web of *N. radiata* is a delicately spun
dome under which the spider hangs and waits for prey (Fig. 5.1). The

(A)

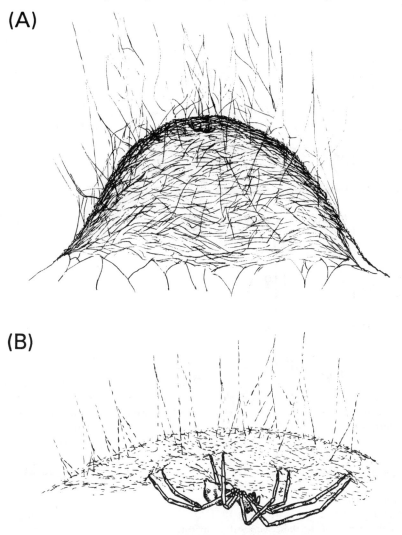

(B)

Fig. 5.1 The linyphiid *Neriene radiata*, the filmy dome spider. (A) Web with spider under the dome. (B) Close-up of spider.

shape of the web and posture of the spider superficially resemble that of the orb weaver *Mecynogea lemniscata*, the basilica spider (Fig. 4.2A). The resemblance reflects convergent evolution, not close kinship. I cleared all the spiders from the bushes and reintroduced spiders at two densities that essentially encompassed the normal range of *N. radiata* densities (Fig. 5.2). Half of the populations at each density were selected at random to

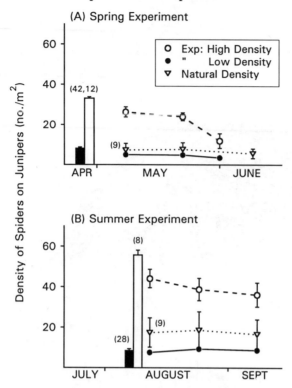

Fig. 5.2 Densities of filmy dome spiders (no./m² of juniper) in the low- and high-density treatments, compared with natural densities on undisturbed bushes. Natural and supplemented prey treatments have been pooled. Means ± 95% CL. Numbers of spiders introduced onto the cleared experimental junipers are represented as open and solid bars (low- and high-density treatments, respectively). The number of junipers in each treatment appears in parentheses. (A) Spring experiment 1973, both sexes combined. Immature spiders were added at the start of the experiment; spiders had matured by the time they were removed. (B) Summer experiment 1973. Only relatively young juveniles were added to the junipers; these were still immature by the end of the experiment. Some mature females immigrated onto the junipers during the experiment; these are included, but always accounted for fewer than 5% of the spiders. (After Wise 1975.)

receive supplemental prey as live fruit flies, which were added separately to each web. The high-density treatments had many fewer bushes because it was not possible to collect sufficient numbers of spiders to increase the number of high-density bushes to that in the low-density treatments.

Casting out immigrants

Failure to compensate for effects of immigrants in the original data analysis produced spurious evidence of density-dependent survival. As a consequence of the experimental design, undetected immigrants had a proportionally greater impact in the low-density populations. The experiments utilized 8–12 junipers with high-density populations and 28–42 similarly sized bushes with low-density populations. Initial densities (no./juniper) were 7 × higher on the high-density bushes. The same number of immigrants would appear on each juniper during the experiment if immigration were independent of the density of resident spiders on the experimental bushes, but the effects of the immigrants would not be evenly distributed among treatments. Because the low-density bushes had fewer spiders, the impact of immigration upon net proportion remaining at the end of the experiment (= survival) would have been greater in low-density than in high-density populations. I chose to ignore this potential problem because I could not rule out the possibility that the actual rate of colonization by immigrating spiders was density dependent due to limited numbers of web sites on the high-density junipers. Furthermore, because estimated rates of immigration were low (Wise 1974, 1975) I decided not to correct for immigration. Below I present results of a reanalysis of the data that shows I was mistaken in ignoring possible immigrants. But first another admission.

Confessions of a pseudoreplicator

The second problem with my original conclusions is the flagrant sacrificial pseudoreplication in the statistical analyses upon which they are based (Hurlbert 1984). I sacrificed a randomized experimental design that consisted of numerous physically interspersed replicate populations by pooling replicate populations for the statistical analysis. Consequently, residual variation between individual spiders instead of residual variation between means of the replicate populations was used for the error terms in my ANOVAs. For similar reasons my use of χ^2 tests on pooled replicate populations to analyze survival was inappropriate.

A new interpretation

I have reanalyzed the results of my experiments with the filmy dome spider (Wise 1975), confining the reanalysis to the 1973 experiments. In this year the design included both prey and density as factors.

Correct analysis of the data confirms my original conclusion that a shortage of prey limits growth and fecundity of *N. radiata*. Supplement-

Table 5.3. *Effects of supplemental prey and density upon growth of immature filmy dome spiders. The rate of growth on each ground juniper has been calculated as W_f/W_i, where W_f is the average weight of the spiders removed from the juniper at the end of the experiment (9 September 1973), and W_i is the average weight of the spiders added to the bush the first few days of August. Two-way ANOVA was performed on log (W_f/W_i)*

Food supply	Density	No. of junipers	Rate of growth (W_f/W_i) (mean ± s.e.)
Natural	Low	14	1.94 ± 0.05
Supplemented	Low	14	2.24 ± 0.04
Natural	High	4	2.03 ± 0.12
Supplemented	High	4	2.67 ± 0.11

Two-way ANOVA

Source	d.f.	Mean square	F	Significance
Food supply	1	0.27431	38.06	< 0.0001[a]
Density	1	0.07553	10.48	0.0028
Food supply × density	1	0.02636	3.66	0.0648
Error	32	0.00721		

Note:
[a] One-tailed.

ing the prey of females that mature in the summer more than doubled fecundity; mean number of eggs per sac was 45.8 ± 4.7 (s.e.) for the prey-supplemented populations, compared with 18.1 ± 2.6 for populations in which spiders were exposed to natural prey levels [$p(t) < 0.001$, d.f. = 10]. Adding prey to webs of the younger instars in the summer significantly increased their growth rate (Table 5.3). However, the weaker effect of food on fecundity in the spring experiment is not statistically significant when the data are correctly analyzed ($p < 0.15$, one-tailed; Table 5.4). The original analysis resulted in a significant F value ($p < 0.05$, one-tailed; Wise 1975) because the inappropriate error term was used. Originally I also concluded that prey shortages affected emigration rates of females in the spring. A more appropriate analysis of survival, i.e. ANOVA of proportions surviving corrected for estimated immigration, reveals that adding prey did not improve net survival of spring females (Table 5.5), nor of summer juveniles (Table 5.6).

The major result of reanalyzing the data is disappearance of all

Table 5.4. *Effects of prey supplementation and density upon fecundity of spring-maturing filmy dome spiders, 1973 experiment (22 April–2 June). Fecundity has been calculated as the mean egg production per female per juniper for females that laid an egg sac after being removed and placed in an isolator. Of the 202 females removed at the end of the experiment, 56% deposited an egg sac*

Food supply	Density	No. of Junipers	Fecundity (mean ± s.e.)
Natural	Low	21	89 ± 6
Supplemented	Low	21	95 ± 7
Natural	High	6	71 ± 5
Supplemented	High	6	83 ± 6

Two-way ANOVA

Source	d.f.	Mean square	F	Significance
Food supply	1	739.72	1.11	0.1496[a]
Density	1	1936.16	2.89	0.0964
Food supply × density	1	96.96	0.14	0.7054
Error	42[b]	669.34		

Notes:
[a] One-tailed.
[b] Error degrees of freedom for the full design = 50. However, on eight of the junipers either there were no spiders at the end of the experiment, or none of the females that were removed laid an egg sac.

evidence that intraspecific competition affects fecundity and net survival of spring-maturing females (Tables 5.4, 5.5). Although most of the treatment differences are consistent with competition, none of the associated F values is statistically significant. The only possible exception is the F for the effect of density on fecundity, which is close ($p < 0.1$) for a two-tailed test, and would be significant at the 0.05 level in a one-tailed test. I decided in my original analysis not to use a one-tailed criterion because one could predict, in addition to competitive effects, positive consequences of increased density for web-building spiders through enhanced prey capture via the 'ricochet' or 'knock-down' effect (Lubin 1974, Rypstra 1979, 1989, Uetz 1985, 1986b, 1989). In fact, the positive effect of density on growth of immatures in the summer could be interpreted in this context (Table 5.3). The ricochet effect should be

Table 5.5. *Effects of supplemental prey and density upon estimated survival of female filmy dome spiders during the spring 1973 experiment. These spiders matured during the experiment. During the summer 1973 experiment, in which only much younger juveniles were added to the bushes, an average of one mature female filmy dome spider colonized each juniper per month from the surrounding vegetation. This value was used to correct for immigrants in estimating the survival rate during the spring experiment because females were at similar stages of development in both spring and summer (Wise 1976). Proportion surviving on a juniper has been estimated as [(final number − 1)/(0.5)(initial number)]. The initial number was multiplied by 0.5 because 50% of the immature spiders added in April were males. Because the populations were open, this estimate of survival reflects losses to the population from both mortality and emigration*

Food supply	Density	No. of junipers	Proportion surviving (mean ± s.e.)
Natural	Low	21	0.57 ± 0.12
Supplemented	Low	21	0.61 ± 0.09
Natural	High	6	0.47 ± 0.05
Supplemented	High	6	0.59 ± 0.09

Two-way ANOVA[a]

Source	d.f.	Mean square	F	Significance
Food supply	1	0.05200	0.26	0.3073[b]
Density	1	0.03427	0.17	0.6826
Food supply × density	1	0.01279	0.06	0.8026
Error	50	0.20253		

Notes:
[a] Variances are heterogeneous $[0.01 < p(F_{max}) < 0.05]$; attempts to transform the data made the variances less homogeneous.
[b] One-tailed.

more pronounced at higher prey densities, which is consistent with the interaction term between density and prey abundance being close to statistical significance ($p = 0.065$). Other explanations are possible. Perhaps at higher densities larger spiders forced smaller ones to emigrate, or ate them, resulting in a higher proportion of larger spiders in the high-density treatment at the end of the experiment (Spiller, pers. comm.). Such speculation is invigorating, but some caution would be advisable. A similar experiment with the same developmental stages the previous

Table 5.6. *Effects of supplemental prey and density upon estimated survival of juvenile filmy dome spiders from 5 August to 9 September 1973. In the previous year juvenile filmy dome spiders colonized empty junipers at an average rate of 1.5 immigrants per month. This value was used to correct the estimate of survival rate. Proportion surviving on a juniper has been estimated as (final number − 1.5)/(initial number). Because the populations were open, this estimate of survival actually reflects losses to the population from mortality and emigration*

Food supply	Density	No. of junipers	Proportion surviving (mean ± s.e.)
Natural	Low	14	0.95 ± 0.06
Supplemented	Low	14	0.92 ± 0.08
Natural	High	4	0.73 ± 0.07
Supplemented	High	4	0.82 ± 0.02

Two-way ANOVA

Sources	d.f.	Mean square	F	Significance
Food supply	1	0.00492	0.08	0.3920[a]
Density	1	0.15069	2.34	0.1362
Food supply × density	1	0.02480	0.38	0.5396
Error	32	0.05864		

Note:
[a] One-tailed.

year, in which density alone was manipulated, provided no evidence of a positive density effect; and in the spring of 1973 effects of density on fecundity, though not statistically significant, were negative.

Using the correct statistical models to reanalyze the data reaffirms some of the original conclusions, yet yields some important changes. The major modification is disappearance of statistically significant evidence for intraspecific competition for prey. Originally I concluded that prey shortages acted as a density-independent limiting factor for juveniles in the summer, but as a density-dependent factor for females in the spring. Now it is clear that for all stages and in all seasons, shortage of prey acted as a density-independent factor for the filmy dome spider.

An exception

Spiller's (1984b) second field experiment uncovered good evidence that exploitative intraspecific competition lowers feeding rates of *Metepeira grinnelli* and *Cyclosa turbinata*. In both spring and summer experiments, maintaining *Cyclosa* densities at a lower level clearly increased the number and total biomass of prey captured per day ($p < 0.05$ for all comparisons; increases ranged from *c.* 25% to 100%). Evidence for exploitative competition within *Metepeira* populations was not as strong. There was no effect in the summer experiment, and in the spring there was a significant effect on number of prey captured ($p < 0.01$), but no effect on total biomass. The experimental design prevented tests of intraspecific density on web height, growth or fecundity, since Spiller intentionally equalized web height and spider size in the plots in order to minimize variation between treatments.

Evidence from several field experiments leads to the conclusion that intraspecific competition for prey is infrequent among web-building spiders. Population dynamic consequences appear to be minimal when exploitative competition has been detected. Explaining the general absence of interspecific competition among web-building spiders thus becomes a question of explaining why spider densities are usually below competitive thresholds. Below I examine evidence for the importance of abiotic mortality factors, natural enemies, dispersal and territoriality in keeping spider densities below levels at which competition for prey is an important interaction.

Abiotic factors

Exposure to wind, moisture and temperature influence habitat selection in many spiders. Rates at which a desert uloborid replaces its orb or moves to a new web site are correlated with wind intensity (Eberhard 1971). Moisture conditions may influence the placement of webs by several orb weavers (e.g. Cherrett 1964, Gillespie 1987, Markezich 1987). The ultimate moisture-limited spiders are those that have abandoned the terrestrial world: the freshwater *Argynota aquatica* and the intertidal spider *Desis marina* (McQueen & McClay 1983, McClay & Hayward 1987). For other spiders flooding can pose problems. Rovner (1987) has shown that nests of wandering spiders function as physical gills, permitting their owners to withstand submergence for many more days than would be possible outside the nest. Rovner argues that flooding may have been one selective factor favoring the evolution of silk nests. Heavy rains can be an important mortality factor for *Agelenopsis aperta*, a desert agelenid (Riechert 1974b). Rainfall possibly kills substantial numbers of early instars of other species, but extensive quantitative data are lacking.

Temperature plays a major role in habitat selection for several species. Female wolf spiders move to areas where warmer temperatures accelerate development of their eggs, which are carried attached to the spinnerets (Nørgaard 1951, Edgar 1971, Kronk & Riechert 1979). Exposure to bright sunlight causes web spinners to orient at an angle that minimizes body temperature (e.g. Pointing 1965, Robinson & Robinson 1978, Biere & Uetz 1981, Suter 1981), or may force them to remain inactive during the hottest time of day (Riechert & Tracy 1975, Tanaka 1991). Exposure to high insolation may cause a web builder to change habitats, e.g. *Micrathena gracilis* abandoning an open pine forest in favor of deciduous forest with a more closed canopy (Hodge 1987a).

Behavioral responses to differences between macro- and microhabitats in physical factors could force spiders into higher densities, increasing the likelihood of competition (e.g. habitat selection by the desert funnel-web spider, *Agelenopsis aperta*; Riechert & Tracy 1975, Riechert 1981). Alternatively, traveling between habitats or suitable web sites could lower overall density through increased exposure to natural enemies. Thus demonstrating the importance of factors such as temperature and moisture in habitat selection provides no compelling evidence that abiotic factors maintain spider densities below competitive levels.

Several investigators have correlated measures of spider activity and abundance with weather factors. Dondale & Binns (1977) found that rainfall and cumulative temperature measured by degree days explained over 85% of the variance in number of spiders captured in a five-year study. Rushton, Topping & Eyre (1987), employing a multivariate ordination procedure to analyze the spider fauna of 54 grassland sites in England, found that wetness of the site was important in determining spider community structure. Rypstra (1986) uncovered a significant correlation between temperature and the density of spiders active on webs in a northern temperate and two tropical forests, though temperature explained less of the variance than vegetation structure and insect activity (Chap. 7). These studies provide useful information on the impact of physical factors on activity and the identities of spiders that may be associated with particular physical factors, but this approach does not indicate the extent to which variation in abiotic factors reduces spider densities. Experimental studies, or correlative studies between physical factors and survival of particular stages, provide the most direct evidence for the role of physical factors in reducing density.

Extreme temperature is the most extensively documented abiotic mortality factor for spiders. Cold winter temperatures can have major impacts upon spider densities. The severe winter of 1978–79 in the eastern United States killed so many overwintering basilica spiders that not only were densities below competitive levels, they were below manipulatable levels. *Mecynogea* was so rare that it was not feasible to set up field experiments testing for competition with *Metepeira labyrinthea*. An unusual winter may have aided the labyrinth spider in escaping competition from *Mecynogea*. The basilica spider is at the northern limit of its range in Maryland, so high overwinter mortality rates may not be surprising. Winter mortality also can be significant for spiders that are widely distributed throughout colder regions. John Martyniuk and I have calculated overwintering mortality rates of 69%, 31% and 25% for filmy dome spiders that were isolated in individual outdoor cages in New York, Maryland and Michigan, respectively (Martyniuk & Wise 1985). Mortality rates were highest among the younger instars. In the autumn Gunnarsson (1987) added sub-adults of another linyphiid, *Pityohyphantes phrygianus*, to spruce branches enclosed by 1-mm-mesh netting, and found overwinter mortality to be 41%, 100% and 60% in three different years. From results of a laboratory study, Gunnarsson (1985) attributed the major cause of disappearance to death by freezing and not predation by other spiders on the branches. He had no direct way to estimate the actual contribution of each mortality factor in the field experiments because individual spiders were not isolated from each other. Male *P. phrygianus* are more susceptible than females to low temperatures; the proportion of males in the population decreased significantly in three of five winters, with the larger decreases occurring in years with colder February temperatures.

A few spiders are active even during winter in cold climates. Some crab spiders and the linyphiid *Pityohyphantes phrygianus* are active on warm winter days in southern Sweden. Several spiders are active under snow, probably feeding mainly on Collembola (Aitchison 1984, 1987). Evolution of winter activity does not imply avoidance of the effects of harsh winters. Adult *Centromerus silvaticus* are active during the winter, yet adults suffer higher winter mortality than the inactive younger instars (Schaefer 1977). Percentage mortality of *C. silvaticus* over the winter is negatively correlated with winter temperatures ($r = -0.86$, $p < 0.05$; one-tailed test).

Thus cold winter temperatures can substantially reduce densities of spiders, affecting not only species at the northern limits of their range, but also species that belong to cold-adapted spider fauna – even spiders that have evolved the ability to remain active during the winter.

Natural enemies

Schoener (1983a), Spiller (1984a) and I (Wise 1983, 1984a) have speculated that in addition to mortality from abiotic factors, losses from natural enemies may contribute significantly to the absence of exploitative competition among spiders. Biologists have compiled an extensive list of natural enemies of spiders (e.g. Bristowe 1941). Accumulating evidence, some derived from field experiments, implicates the importance of birds, lizards, wasps, parasites, parasitoids and other spiders in determining the average abundance of many spider species. Simply listing natural enemies, however, will not prove that they are responsible for keeping spider densities below competitive levels. We require evidence that (1) natural enemies limit spider population densities, and, furthermore, that (2) competition would occur among spiders if control of spider populations by natural enemies were removed. The second condition has yet to be satisfied, but good support exists for the first.

Wasps

In explaining the discrepancy between his discovery of evidence for interspecific competition and the results of other field experiments with spiders, Spiller (1984a) suggests that *Metepeira* and *Cyclosa* were unusually abundant on his sites because of the scarcity of several spider enemies, particularly pompilid and sphecid wasps. He observed that these spiders were less dense in regions of the salt marsh in which wasps appeared to be more abundant. Spiller suggests wasps were absent from his study site because the ground was too hard for nesting. In contrast, ground-nesting vespid wasps were abundant in the forest in which we manipulated densities of *Metepeira labyrinthea* and *Mecynogea lemniscata*. We did not sample quantitatively, but judged the wasps to be at least moderately abundant from the frequency with which we accidentally sampled their nests. In 1980 the vespids were more than a minor irritant, as they threatened not only to prevent spider competition, but also the planned competition experiment.

As a first step in setting up the experiment with adult female labyrinth spiders (Wise 1983), we had collected over 250 spiders, measured their size and marked them uniquely on the abdomen with dots of colored paint. We then added them to open experimental frames and allowed them to build webs. We felt comfortable releasing such precious specimens because in an experiment conducted earlier in the summer over 80% of the juvenile spiders added to these frames remained and built webs. Furthermore, colonization rates of marked females were high in a pilot experiment performed the previous year (1979). To our dismay, colonization of marked adult females in 1980 was much lower than expected, and on one high-density unit was an order of magnitude lower than on the others. Observations over the next few days revealed that vespids were removing *Metepeira* from their webs. We also discovered a large vespid nest near the one particularly ill-fated experimental unit. After this nest had been destroyed, colonization of the unit dramatically improved the next time we added spiders (Table 5.7). In previous experiments (Wise 1979, 1981) *Metepeira* survival on the units was lower during the first part of August than during subsequent weeks. Originally I had attributed this initial poor survival to high rates of emigration of spiders that had not yet located suitable web sites. However, the data on the effects of vespids lead to the hypothesis that they are a significant mortality factor for adult *Metepeira*.

This evidence should be interpreted with caution. First, the vespid manipulation was not replicated, nor was there a contemporaneous control, primarily because the effect of vespid predation was not anticipated. Improvement of survival on the units could have been merely coincidental with destruction of the vespid nest. Also, in order for the findings of 1980 to be consistent with the absence of such dramatic rates of disappearance in earlier years, one would have to hypothesize that vespid feeding on labyrinth spiders is intense only for a brief time, or that this one colony was particularly large, or particularly effective in finding and capturing *Metepeira*, especially spiders on the third high-density unit. Nevertheless, it is difficult to refrain from speculating that vespids are an important source of mortality for *Metepeira*, perhaps exhibiting a behavioral response to dense aggregations.

Do Spiller's and my findings together constitute good evidence for the importance of wasps in preventing competition in *Metepeira* populations? The evidence is suggestive, but on balance must be judged speculative. My manipulation implicates the importance of vespid predation, but it is not known if *Metepeira* would have been at competitive densities if wasp predation had been curtailed earlier in the season, perhaps during the last half of July. It is risky to compare Spiller's research on *Cyclosa* and *Metepeira grinnelli* with my studies of *Mecynogea* and *Metepeira labyrinthea* because we did not measure densities of spider natural enemies nor prey abundances. Although spider densities were

Table 5.7. *Survival of marked labyrinth spiders,* Metepeira labyrinthea, *on open experimental frames before, and after, a large colony of vespid wasps had been intentionally destroyed*

Treatment	No. added	Number (%) present after 2–3 days	Number (%) surviving from Day 0			
		Day 0	Day 3	Day 4	Day 5
Vespids present; Spiders added 30 Jul.–1 Aug.						
Low density[a]	100	57 (57%)	30 (53%)	28 (49%)	24 (42%)
High density						
Unit 1	48	21 (44%)	16 (76%)	15 (71%)	15 (71%)
2	48	17 (35%)	9	9	9
			(53%)	(53%)	(53%)
3	75[b]	4 (5%)	0 (0%)	0 (0%)	0 (0%)
Vespids absent; Spiders added 20 August						
Low density	83	54/72[c] (65%)	70 (97%)	70 (97%)	70 (97%)

response would not be expected, given the low number of eggs laid per female (c. 10). Wasps may play a role not only in keeping populations of many spiders at low levels, but may also help regulate spider population densities if they forage in a strongly density-dependent manner. Clearly more research into this area is warranted, and should prove fruitful.

Documentation of actual effects of wasps on spider populations appears to be limited. Laing (1979) estimated mortality rates from pompilids to be 12–30% for an Australian funnel-web spider, and hypothesized that the major impact of such predation is to reduce intraspecific competition. He points out that his estimates are crude because he could not precisely identify all causes of disappearance during the summer. Laing discovered that the younger spiders are immune from attack because their tunnels are too small for the wasp to enter. McQueen (1978, 1979) found that the immature instars of the burrowing wolf spider *Geolycosa domifex* also were relatively safe from predation by a pompilid wasp because they closed their burrows during times of wasp activity. The wasps successfully attacked almost all the mature females when the latter left their burrows open so that newly hatched young could emerge and disperse. Because of this timing, the impact of wasp predation on the density of the spider population is minor. Conley (1985) discovered that rates of parasitism by the pompilid wasp *Paracyphononyx funereus* were high (c. 50%) among adult female burrowing wolf spiders in the autumn; summer rates, however, were only c. 5%. Thus it appears that mortality from wasps in this burrowing lycosid also is most intense among post-reproductive females. These findings contrast with my observation that vespid wasps potentially exert high mortality on female labyrinth spiders during the height of their reproductive period. Conley's and McQueen's studies suggest that pompilid predation is not a major limiting factor for populations of burrowing wolf spiders.

Fincke, Higgins & Rojas (1990) monitored rates of parasitism of the orb weaver *Nephila clavipes* by an ichneumonid ectoparasite on Barro Colorado Island, Panama. Intermediate-sized juvenile females were parasitized most heavily; rates were between 25 and 30% in two consecutive years. The investigators report that this particular species of ichneumonid had not been noticed in previous studies of *N. clavipes* on Barro Colorado Island, leading them to propose that an outbreak of this parasite contributed to an apparent decline in the *N. clavipes* population during the second year of their study.

Spiders

Spiders can be their own worst enemies. Members of the Family Mimetidae, which prey only on other spiders, are the most notorious. Kleptoparasites of the Genus *Argyrodes* (Theridiidae) steal prey from the webs of other spiders, but if the host is small enough, *Argyrodes* may forsake thievery and eat the host (Archer 1946, Lamore 1958, Exline & Levi 1962, Smith-Trail 1980, Wise 1982, Tanaka 1984, Larcher & Wise 1985). *Argyrodes fissifrans* inhabits webs of the agelenid *Agelena limbata* and attacks the much larger host during, or just after, molting (Tanaka 1984). Some salticid species invade spider webs and may steal prey, eat the host or consume the host's eggs (Jackson & Hallas 1986a,b, Jackson 1988). Spiders frequently comprise a substantial fraction of the diet of wolf spiders (Edgar 1969, Hallander 1970a). These observations implicate predation by other spiders as an important limiting factor, but firm proof comes only from field experiments.

Field experiments have demonstrated the impact of spider predation on the density of other spiders. Experiments designed to test for interspecific competition among wolf spiders provided evidence that interspecific predation affected the population density of one species (Schaefer 1974, 1975; details discussed in Chap. 4). By adding mature *Argyrodes trigonum* to open frames on which *Metepeira labyrinthea* had established webs, I demonstrated that this kleptoparasitic theridiid readily invades webs occupied by the labyrinth spider, producing a substantial decline in density by either forcing emigration or by eating the ones that linger too long (Wise 1982; Fig. 5.3). Additional field experiments confirmed these observations for *Metepeira* and also demonstrated that *Argyrodes* is a mortality factor for the filmy dome spider (Larcher & Wise 1985).

Overwintering spiders on spruce branches may consume some of their smaller arachnid neighbors before spring arrives. Gunnarsson (1985) discovered that the majority of winter-active spiders collected from spruce were the linyphiid *Pityohyphantes phrygianus* (40%) and crab spiders of the genus *Philodromus* (23%). Both of these spiders consumed a substantial fraction of the smaller spiders with which they were confined in laboratory studies conducted at 4 °C. Rates of intraspecific predation were much lower among *Pityohyphantes* and *Philodromus*, and among the smaller spiders themselves. In an earlier field experiment Gunnarsson (1983) had compared overwinter mortality among spiders at two

Table 5.8. *Rates of disappearance of web spinners in each of four experimental treatments: TR 1 – Unmarked spiders on open vegetation; TR 2 – Marked spiders on open vegetation; TR 3 – Marked spiders in topless cages; TR 4 – Marked spiders in complete cages. Cages were 40 × 30 × 30 cm. Only spiders that had started web construction on the open vegetation or in a cage are included in the analysis. Cages in TR 4 permitted passage of the spider, its prey and invertebrate natural enemies, but excluded vertebrate enemies. Rates coded with the same letter do not differ significantly at the 0.05 level according to a G-test*

Forest site/time period	n	Percentage disappeared			
		TR 1	TR 2	TR 3	TR 4
Temperate (USA)					
0300–0900	60	23 a	20 a	17 a	15 a
0900–1500	60	28 a	32 a	25 a	20 a
1500–2100	60	30 a	32 a	23 a	17 a
2100–0300	60	23 a	18 a	27 a	22 a
Sub-tropical (Peru)					
0900–1500	60	65 b	68 b	58 b	25 a
2100–0300	60	33 a	38 a	30 a	18 a
Tropical (Gabon)					
0900–1500	30	60 b	50 b	53 b	20 a
2100–0300	30	17 a	27 a	23 a	10 a

Source: Rypstra 1984.

density of large spiders on spruce branches in September and October, suggesting that spiders are a limited food resource for birds.

Predation by birds may also exert significant pressure on web-building spiders in tropical regions. Rypstra (1984) discovered that survival of spiders in tropical and sub-tropical areas was higher when they were protected by cages that excluded vertebrates but did not prevent movement of spiders and their prey. The cage had no effect upon survival in a northern temperate forest, nor at night in the tropics (Table 5.8). Rypstra suggests that the cages may have reduced predation not only from birds, but also from primates and large dragonflies; she speculates that pressure from daytime predators may explain why web spiders in her Peru and Gabon sites are predominantly nocturnal. Horton (1980) and Eisner & Nowicki (1983) obtained experimental evidence that the zig-zag stabilimentum, which is woven into the orb by many tropical and temperate spiders, reduces losses from bird predation.

An exclosure experiment has provided convincing evidence that birds consume a substantial fraction of a population of the funnel-web spider, *Agelenopsis aperta* (Agelenidae), that inhabits an Arizona riparian woodland (Riechert & Hedrick 1990). Preliminary studies implicated birds as a major mortality factor. Several species were observed capturing *A. aperta*, and close monitoring of marked spiders in a fenced 900 m² area revealed that 63% and 72% disappeared in a six-week period in two successive years, respectively. Riechert & Hedrick were able to eliminate migration, death due to fighting with conspecifics, and death during prey capture or molting as causes of the documented *A. aperta* disappearances. Confirmation of the importance of avian predation comes from a field experiment conducted over two years. The investigators compared the rates of disappearance of *A. aperta*, over a period of seven days, from paired adjacent 10 m² plots, one of which had been covered with bird-netting. Riechert & Hedrick replicated this manipulation 10 times in one season and 8 times the next year. In both years survival of the protected spiders was higher (Fig. 5.4).

Riechert & Hedrick conducted a similar experiment (size of plots was modified because of lower spider density) with *A. aperta* inhabiting a desert grassland site. Three years of experimentation uncovered no significant differences in survival rates between exposed web sites and those protected from bird predation. Lack of an effect corresponds with the low incidence of bird predation observed in the grassland habitat (one event in 20 years). This difference in predation intensity between grassland and woodland is correlated with genetic differences in antipredator behavior between the two populations (Riechert & Hedrick 1990). Second-generation descendants reared in the laboratory displayed differences in the time elapsed before resuming their standard foraging stance after having retreated to the funnel in response to a web disturbance meant to mimic the approach of a predator. Spiders from the riparian population, which is exposed to greater predation pressure from birds, remained in the retreat longer before resuming their position at the entrance of the funnel (1195 ± 148 s versus 82 ± 6 s, $p(U) < 0.0001$). Also, a greater fraction of the riparian spiders retreated following the initial stimulus.

This experiment, like those of the Swedish researchers, cannot unambiguously finger birds as the agent of predation. Nevertheless, natural history observations, in conjunction with the clearly improved

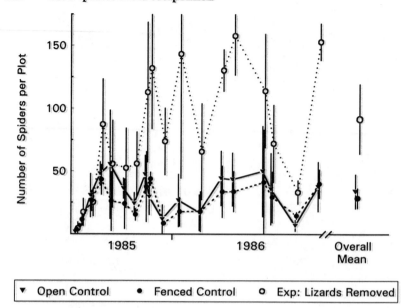

Fig. 5.5 Impact of lizards upon population densities of web-building spiders. Mean numbers (± 1 s.d.) of web spinners (all species combined) in three lizard-removal enclosures, three fenced control plots, and three unenclosed plots. Lizards were removed throughout the experiment. The first census was 11 May 1985, the last was 19 December 1986. Overall means are based upon numbers at each census in a particular plot from 17 June 1985 through 19 December 1986. (After Spiller & Schoener 1988.)

rate in the fenced lizard-removal plots, even though prey was less abundant in these plots than in the open control plots, which had normal lizard densities. Spiller & Schoener (1988) speculate that lizards may exclude spiders from areas where prey abundance is higher (by 'direct predation or interference') and/or spiders may not forage outside their retreats as frequently in areas with lizards. They conclude that '. . . predation appears to be the major interaction responsible for the lizard effect.'

Spiller & Schoener discovered that removing lizards increased the average number of spider species per plot by 1.3 (3.2 versus 2.5). Since lizard predation did not promote spider coexistence, interspecific competition between spiders appears to be unimportant. In further support of this conclusion, Spiller & Schoener (1988) point out that overall spider densities in their study area were similar to those in studies failing to detect competition between orb weavers. Absence of competitive densities may not be due primarily to lizard predation. Periodic extremes of weather may help keep spiders at densities below competitive levels in this community. Schoener & Toft (1983a) report the contemporaneous population explosion of *Argiope argentata* (from 0 to *c.* 115) on six different islands in less than a year, which suggests an important role for physical factors in setting spider abundance in this island system. Although *Metepeira datona* densities do not fluctuate as much as those of *A. argentata* (Fig. 33.6 in Schoener 1986), weather may still play an important role in limiting densities of *M. datona* in Schoener & Spiller's study area. They attribute two sudden declines in spider abundance during the second half of their experiment to severe winter weather and a hurricane.

By removing lizards for 1.5 years, Spiller & Schoener clearly demonstrated that lizards affect numbers of web-spinning spiders. Yet because some questions remained unanswered, particularly with respect to the possible importance of interference and exploitative competition, the investigators decided to continue the study for another 18 months. Three years of experimentation revealed that both predation and exploitative competition are important in this system (Spiller & Schoener 1990a). The evidence is discussed in detail in Chapter 8, where the problem of uncovering direct and indirect effects in guilds of generalist predators is discussed in detail.

Schoener (1986) has experimented directly with populations on small islands in order to demonstrate the effect of lizard predation on spider densities. He added three females and two males of *Metepeira datona* to each of 15 small islands, none of which had this orb weaver, and only five of which had lizards. The introduced *Metepeira* had become established on four of the lizard-free islands after a year, but on only one of the five islands inhabited by lizards.

Pacala & Roughgarden (1984) also obtained clear experimental proof that lizards depress densities of orb-weaving spiders. On an island in the Antilles they constructed six fenced 12 × 12 m plots and then removed two species of *Anolis* lizards from all six plots. Lizards were reintroduced to three of the plots, resulting in densities *c.* 3 × higher than in the other three fenced areas. The density of lizards in the high-density treatment was similar to that in an adjacent 160 m² open area. After six months the numbers of orb weavers (total of the three most conspicuous species) were 1, 1 and 4 in the high-density plots, and 20, 36 and 33 in plots with the lower density of lizards. This remarkable difference between treatments is most reasonably attributed to differences in rates of

not necessarily closely linked to negative effects of competition among older stages.

Territoriality

In contrast to early dispersal behavior, territoriality is an evolved behavior that is tightly linked to competition among adults. Susan Riechert has proposed widespread territorial behavior as an explanation for the rarity of exploitative competition for prey among spiders. She bases her theory upon extensive studies of the desert funnel-web spider *Agelenopsis aperta*, which actively defends territories larger than the web. Average territory size is large enough to ensure that spider densities are below levels at which exploitative competition for prey would occur. Riechert has studied this system for over a decade and has written several comprehensive reviews of her research (Riechert 1982, 1986, 1988, Riechert & Gillespie 1986).

Much of Riechert's research has been with an *A. aperta* population in central New Mexico, where the species inhabits a physically harsh desert grassland growing on an old lava bed. Suitable web sites must be sufficiently shaded to allow the spider enough daylight hours to be active on the sheet to capture prey; high-quality web sites also are associated with vegetation that increases insect abundance. The population exhibits a clumped dispersion pattern because areas with suitable web sites are scattered. Within clumps *A. aperta*'s pattern is overdispersed because spiders defend their webs and a surrounding area against intruders, preventing other *A. aperta* from building a web within this territory. The population consists of territory owners and floaters without web or territory, who wander in search of a suitable site or an occupied territory which they attempt to wrest from the owner. Larger spiders tend to win aggressive encounters, but there also is a bias favoring the owner. The intensity of territory defense is a function of web-site quality measured directly in terms of prey capture rate, growth and fecundity of the spider. In an ingenious experiment, Riechert placed two sticky-trap models of an *A. aperta* web within territories from which the owner had been removed, and compared the capture rate of these traps with single traps in other vacant territories. Insect capture rates were 60% lower in the presence of another trap, clearly demonstrating the advantage gained by preventing another spider from building a web within a territory.

Further proof that the territorial system of *A. aperta* is energy based comes from interpopulation differences in territory size that correspond to differences in prey availability. In a physically more benign and more productive riparian habitat in Arizona, average territory sizes are 0.1 m² versus 4–10 m² in the desert grassland in New Mexico (Riechert 1978a). In the Arizona habitat *A. aperta* tolerate webs above them, as do many families of web-spinning spiders that live in meadows and deciduous forests, where suitable web sites are not as limited as they are for *A. aperta* in central New Mexico. Fights over territories between spiders from the riparian population are less costly in terms of number of different acts, duration of the encounter and risk of injury, than are territorial disputes among grassland *A. aperta*. Differences between Arizona and New Mexico spiders in territory size and the intensity of agonistic behavior result from genetic differences between the populations in both sex-linked and autosomal loci. These two populations of *A. aperta* have adapted to different levels of prey abundance by evolving different territorial behaviors.

Despite the correlation between populations in territory size and prey abundance, within the New Mexico population territory size does not change in response to temporal fluctuations in insect availability. *A. aperta* has evolved a territory size that prevents exploitative competition for prey in the years of lowest insect abundance. Apparently changes in prey abundance at a site within a season are too brief and erratic to have favored a flexible behavioral phenotype that would respond quickly to changes in insect availability. This discovery, in combination with other patterns uncovered by Riechert, establishes that the evolution of territoriality in *A. aperta* has been driven by selection favoring spiders that own the most productive territories, which not only contain the best web sites but are large enough to prevent exploitative competition for a limited food supply. Intense, direct competition for territories has eliminated exploitative competition for food among this agelenid.

Riechert has argued that territoriality is widespread among spiders (e.g. Riechert 1978a, Riechert & Lockley 1984, Riechert & Gillespie 1986). If this hypothesis is true, the absence of significant intra- and interspecific competition for a limited prey supply would be a direct consequence of behavioral interactions between spiders of the same species. This is potentially a far-reaching generalization, because it would explain the absence of competition without invoking niche partitioning or the impact on population density of natural enemies and abiotic mortality factors. It is an hypothesis that agrees with the aggressive behavior of a generalist predator that readily defends its web against conspecific invaders. Biologists in addition to Riechert believe

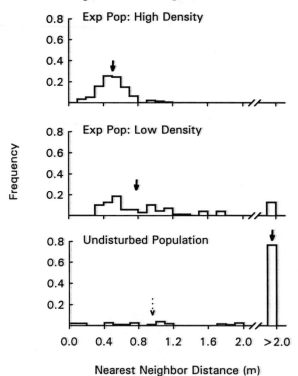

Fig. 5.6 Relative degree of crowding experienced by adult female *Metepeira labyrinthea* in the high- and low-density experimental populations on open frames, and on natural vegetation in a nearby, undisturbed area. Crowding is measured by the distribution of nearest-neighbor distances, measured on Day 0 (21 August) of the experiment for the manipulated populations, and 6 September for the undisturbed area. Solid arrows mark the median of each distribution; the dotted arrow indicates the median of all nearest-neighbor distances ≤2 m in the non-experimental population. (After Wise 1983.)

the presence of floaters would be insufficient to prove the existence of an energy-based territorial system. Floaters could be individuals that were unsuccessful in obtaining a web site in a habitat in which the supply of suitable sites is limited. Conceivably these sites might be so scattered that exploitative competition for prey is absent. Such a system, in which spiders compete directly for web sites, would resemble the *A. aperta* system in many ways but would not constitute prey-based territoriality.

Territoriality in *A. aperta* possibly is an extreme behavior that reflects the environment in which this particular agelenid lives. Interactions between vegetation structure, the qualities of exposed lava, and the extreme heat and dryness of central New Mexico result in a limited supply of areas that will support growth and reproduction of *A. aperta*. Within these high-quality areas *A. aperta* can defend a territory in excess of the web because, although it is a web builder, *A. aperta* is closely related to wolf spiders and hence is well adapted to patrolling areas off the web in search of intruders who have built webs in its territory. Web builders inhabiting other habitats may find suitable web sites much more abundant, and non-agelenid spiders may have greater difficulty, and be more exposed to predation, exploring vegetation off the web on a frequent basis. Territoriality of the type so exquisitely explored by Riechert in populations of *A. aperta* may not be widespread among web-building spiders. Her hypothesis that resource-based territoriality is widespread among spiders clearly merits intensive testing with a variety of species.

Web-weaving myopia

Web spinners have poor eyesight yet can maneuver nimbly on their webs, but myopic ecologists may become tangled in their own spinning work if their vision is too nearsighted. It is risky to generalize about the general spider persona from results of field experiments with web builders. Almost all field experimentation into resource limitation, competition and territoriality has been conducted with spiders that forage by using webs to trap prey. In addition to the role of natural enemies, abiotic factors and early dispersal in keeping spider densities low, the foraging mode of web building may limit the intensity of competition for prey. Exploitative competition may be rare among such spiders because by sampling an aerial insect plankton a web builder may have minimal immediate impact upon neighboring spiders unless the position of the web is such that it interferes with the access to prey of close neighbors. Because of their sedentary nature and the physics of prey capture, web builders exert immediate negative effects upon neighbors only if their webs are exceptionally close. Furthermore, feeding by a web builder may have minimal long-term impacts on more-distant neigh-bors, or on members of other species that are not immediately adjacent, because the source of prey is not localized to the habitat occupied by the web-builder population. Effects of prey shortages upon most web-spinning spiders may be largely density independent because a particular population of web-building spiders exerts only a minor impact upon the rates of growth of its prey populations. Thus the widely different

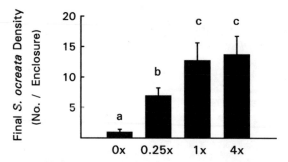

Fig. 5.8 Nearly complete convergence of *Schizocosa ocreata* densities. Total numbers of spiders removed by trapping and searching the enclosed plots 2.5 months after spiderlings had been added to the enclosures. Densities identified with the same letter do not differ significantly from each other. Those with different letters differ at the 0.05 level by the SNK test. (After Wise & Wagner 1992.)

more *S. ocreata* (Fig. 5.7B). Collembola, or springtails, are abundant insects in the leaf litter and upper soil. They are likely major prey of small wolf spiders. Taken together, the negative effect of spider density on growth rate and this nearly significant negative effect of spider density on Collembola numbers suggest that *S. ocreata* spiderlings depressed Collembola populations, and as a consequence experienced exploitative competition.

At the end of the experiment, Collembola numbers did not differ between density treatments. A possible explanation is the nearly complete convergence of spider numbers in the $0.25 \times$, $1 \times$ and $4 \times$ treatments (Fig. 5.8). Cannibalism, density-dependent mortality from natural enemies and emigration all probably contributed to the convergence of population densities. It is noteworthy that any effects of competition on growth rate were much less than the dramatic effects of elevated density on the proportion of spiders persisting to the end of the experiment. The mechanisms underlying this convergence in densities of young *S. ocreata* merit further investigation.

Synopsis

Field experiments have shown that prey shortages act as a density-independent limiting factor for web-spinning spiders. Therefore, web builders do not exhibit exploitative interspecific competition for prey because intraspecific competition is absent or weak. In an experiment with old-field orb weavers, negative intraspecific effects did not appear at all during the second year, when densities were higher and prey appeared to be limiting growth more than the previous year. Intraspecific density effects in an experiment with the basilica and labyrinth spiders were minor. Originally I attributed the absence of intraspecific competition in this system to a superabundance of food in the year the experiment was conducted. However, field experiments in following years with the labyrinth spider clearly demonstrated that intraspecific competition is absent even when prey shortages limit growth and fecundity.

Earlier experiments with the filmy dome spider led me to conclude that prey shortages acted as a density-independent factor for immatures, but that competition for prey among adult females lowered fecundity and increased emigration. Discovery of intraspecific competition was a major motivation for designing future experiments with other web spiders to test the hypothesis of interspecific competition for prey. Re-examination of the results of my earlier studies reveals that the apparent negative density effects among female filmy dome spiders were artifacts of not adequately compensating for immigration into the open populations, and a statistical analysis that was sacrificially pseudoreplicated. Reanalysis of the data with the appropriate models reveals that filmy dome spiders were food limited, as was originally concluded, but that intraspecific competition was absent. Thus these results agree with later experimental evidence for the lack of importance of intraspecific competition for food among web-building spiders.

Explaining how spiders avoid interspecific competition becomes a question of explaining why spider densities frequently are below competitive levels. Abiotic mortality and natural enemies may have substantial impacts on many spider populations. Conditions of moisture and temperature influence habitat selection in many species and probably cause considerable mortality in some species even within preferred habitats. Extreme cold is the most extensively documented abiotic mortality factor in spiders; it has been shown to kill a large fraction of several species in some years. A wide spectrum of parasites, parasitoids, and invertebrate and vertebrate predators attacks spiders.

Predation by vespid wasps appears to have depressed densities of adult female labyrinth spiders in experiments in which competition was absent; Spiller speculates that the absence of wasp predation in his study area may have permitted spider densities to reach levels at which competition was important. Evidence is fragmentary, but it does implicate wasps as major sources of mortality in some spider populations. In addition to wasps and parasitoids, spiders themselves are major

importance. Mark Trail, unaging and wise naturalist of the Sunday comics, explains that spiders are 'beneficial to all of us . . . the harmful pests they destroy in a small section of land mounts up to an astronomical figure' (*The Washington Post*, a Sunday in 1988). Popular culture's perception of the spider's role in the widely accepted balance of nature parallels a scientific folklore that also elevates spiders to the hero's pedestal. Early in this century several investigators estimated the densities of spiders in assorted terrestrial ecosystems (Bristowe 1939). Perhaps the most imaginative estimate is 11 000 spiders per acre of woodland, derived from a count of nine spiders in a 4 square foot patch of forest. Bristowe (1971) claims that during some seasons a particular field in Sussex had over 2 000 000 spiders per acre. A few quick calculations lead him to conclude that the weight of insects consumed by the entire British spider fauna in a year exceeds the combined weight of all the humans in Britain. The full implications of this finding are yet to be determined.

Because spiders are 'usually a major component of the predator biomass in most terrestrial ecosystems,' Hagstrum (1970) argues that spiders 'consequently are important biological control agents.' After reviewing the literature, Kajak (1965) concludes that '. . . in many situations the spiders may play an important role as one of the significant factors controlling the number of insects.' Dabrowska-Prot, Łuczak & Tarwid (1968) cite several authors who advocate a similar role for the spider persona in the ecological dramas that unfold on several terrestrial stages. Nyffeler & Benz (1987), however, are more cautious in assessing the evidence gathered to date. They conclude that 'the significance of . . . (spiders) . . . as natural control agents is still largely unknown.'

Reasons to predict an impact of spiders on insect populations

Spider predation has been intense enough to mold the evolution of prey characteristics: predation by salticids has shaped the morphology and behavior of some tephritid flies. Their wing markings resemble the pattern of the legs of jumping spiders; the flies also wave their wings in a fashion that appears to mimic the agonistic behavior of salticids – making them 'proverbial sheep in wolf's clothing' (Greene, Orsak & Whitman 1987, Mather & Roitberg 1987).

Negative correlations between spider abundance and prey numbers could be cited to support the claim that spiders have substantial impacts upon densities of their prey. For example, Dabrowska-Prot & Łuczak (1968a) discovered that the abundance of mosquitoes in an alder forest declined as the densities of three species of web-building spiders increased steadily over the season. Correlations of course do not establish causal connections, but patterns such as this one are consistent with estimates of energy flow to spider populations, which suggest that spiders substantially affect prey densities.

Food chains and energy flow

Studies of energy flow and the movement of labeled materials through food webs indicate that spiders are major components of the predatory fauna, and, more importantly, that spiders capture a substantial fraction of the insects in the trophic levels beneath them.

Spiders clearly comprise a major share of the invertebrate predators in many terrestrial ecosystems. Van Hook (1971) concluded that wolf spiders (*Lycosa* spp.) were the major invertebrate predators in an eastern North American grassland, although their fraction of the predatory species is unknown because he gives no estimates of densities of other predators. Spiders are major predators on the forest floor (van der Drift 1951, Reichle & Crossley 1965, Norton 1973, Nyffeler 1982). Moulder & Reichle (1972) discovered that spiders were 2.7× as numerous as centipedes or predacious beetles in a ^{137}Cs-labeled tulip poplar (*Lirioden-dron*) forest, and had 25% more biomass than either group. Manley, Butcher & Zabik (1976) employed labeled prey to establish the importance of spiders among the top arthropod predators on the floor of a beech–oak forest. In this study DDT-resistant Collembola were fed DDT-contaminated yeast before being introduced into fenced field plots. Analysis of the concentrations of a break-down product of DDT among the predators revealed that spiders consumed 55% of all the Collembola eaten by top arthropod predators; centipedes were the next most important predators on this important insect group. Although results such as these are germane to predicting the importance of spiders relative to other top arthropod predators, such information does not indicate what fraction of the prey populations the spiders consume.

Good evidence has been collected to establish that the spider assemblage grabs a large fraction of insect production. Moulder & Reichle (1972) obtained quantitative estimates of the impact of spiders in the forest-floor community they studied. Rates of consumption were estimated from information on concentrations and rates of turnover of radioactive cesium in the litter and in components of the fauna. Spiders consumed 44% of the mean annual standing crop of all cryptozoans on the forest floor. This figure underestimates the potential effect of spiders,

sis that spider predation limits insect populations, because the manipulations did not substantially alter spider densities. After two weeks the density of spiders in the cages did not differ between treatments for either experiment. Within the cages substantial predation by spiders upon other spiders appears to have influenced the extent to which the spider complex affected prey numbers. Even removing spiders appeared to have increased the survival of small spiders enough to compensate partially for the initial reduction in spiders. Breymeyer concludes from her study that cannibalism is a significant factor controlling the densities of wandering spiders. The results of her experiments support this conclusion more than they argue for prey limitation by spiders, though even with this conclusion one must be cautious because of possible cage effects. Breymeyer acknowledges that her studies are preliminary.

Western black widow spiders (Theridiidae) that build webs near the nests of harvester ants eat foraging workers that become tangled in the snares. MacKay (1982) studied five ant nests among which the number of spiders per nest ranged from 2 to 15. Widow spiders began to appear at the nests in June. As the number of spider webs increased the foraging activity of the ants declined, until after a few weeks it had ceased completely in four of the nests. At the fifth nest foraging continued all season, but this nest always had only a few spiders. MacKay removed the spiders from the vicinity of the four inactive nests, and foraging commenced within 24 hours in three of them, in two cases attaining rates twice that observed earlier in the season (Fig. 6.1). This experiment had no contemporaneous controls, but the pattern before the manipulation and the rapidity of the response to the spider removal certainly suggest that the spiders had a major impact upon the ant behavior. It appears, though, that the widow spiders probably exert no noticeable impact upon ant populations. MacKay calculated that a nest loses from 2 to 30 ants per day to spider predation, which represents only about 0.2% of the colony. Seven to ten days after the ants became inactive, the spiders abandoned their webs and the ants resumed foraging. Measured over the life span of the colony, mortality and the interruption in foraging due to widow spiders are negligible.

These two studies do not provide convincing evidence that spiders exert major impacts upon their prey populations. MacKay's study provides clear evidence of an effect, but not one that has major population-level consequences for the prey. Breymeyer's study failed to establish adequate differences in spider densities. The scope of both studies was limited, either because only one spider and its prey were

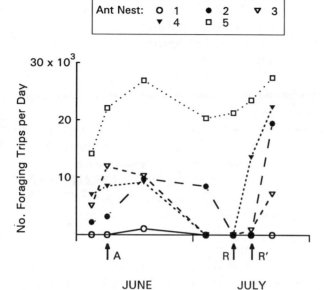

Fig. 6.1 Effect of predation by the western widow spider upon foraging activity of harvester ants. Activity for each of five nests is shown. Active webs were first noticed near the nests in early June (arrow A). Arrows R and R′ indicate the dates that all spiders were removed from nests 1 and 4, and 2 and 3, respectively. Nest 5 had few spiders throughout the season and no manipulation of spider numbers was done. (After MacKay 1982.)

studied, or because experimental units were small and the manipulation lasted a relatively brief time. Thus the absence of strong effects of spiders in these experiments does not argue persuasively against the importance of spiders in the dynamics of their prey.

A classic study

Clarke & Grant's (1968) study of spider predation in a forest-floor community is the most widely cited field experiment designed to uncover the role of spiders as predators. Riechert & Lockley (1984) state that Clarke & Grant (1968) were the first to demonstrate experimentally 'that spiders can have a strong stabilizing influence on prey.' This study is cited in recent reviews in support of the conclusion that spiders contribute significantly to the control of insect populations inhabiting soil/litter systems (Nyffeler 1982, Nyffeler & Benz 1987).

Clarke & Grant located four 13 m² areas 'chosen for structural uniformity' in a beech–maple forest. They removed as many spiders as possible from one area, which was enclosed with a sheet-metal fence, by sieving litter over a one-week period. Another fenced area in which litter was sieved but spiders were not removed served as a control. Two open areas served as controls for the effect of enclosing the plots. One of these open controls was undisturbed, and litter was sieved in the other. Plots were sampled by taking ten 0.09 m² samples from each area on several sampling days over a 10-week period. Each plot had been sampled once before the week of the perturbation. Over the course of the study the average number of spiders per sample in the removal plot was approximately half the number in the three control plots; numbers of Collembola, a major prey of spiders, were highest in the removal plot (Fig. 6.2). Densities of centipedes, a major predatory group of the forest floor that is also preyed upon by spiders, were also highest in the removal plot (Fig. 6.3A). The researchers determined in laboratory studies that the spiders in this community did not prey upon millipedes. Densities of millipedes in the removal plot were intermediate between those of the control plots (Fig. 6.3B). The directions of the differences and the internal consistency of the patterns led Clarke & Grant to conclude that 'predation by spiders is an important subtractive process acting on populations of centipedes and collembola.'

Unfortunately, flaws in experimental design and inappropriate statistical analyses make such a strong conclusion suspect. Clarke & Grant recognized the design shortcoming – that the experiment was not replicated. They point out that 'it is possible that the abundance of prey in (the removal plot) would have been observed whether spiders were removed or not. Replication of the various experimental treatments would have allowed us to eliminate this possibility.' They attempt to circumvent this problem partially by looking at the internal consistency of the results, and also by using statistics to show that centipede numbers were significantly different between the removal and fenced control plots. However, this analysis is doubly pseudoreplicated (Hurlbert 1984), because samples were pooled within single experimental units on one sampling date and also across the entire 10-week sampling period. The resulting t-statistic is statistically significant ($p < 0.005$, d.f. $= 79$), but is irrelevant to a test of the hypothesis that spiders influence centipede numbers. Any difference between removal and control plots unrelated to the spider manipulation would have entered the calculation of the t-statistic so many times as to make it meaningless as a test of the hypothesis under scrutiny.

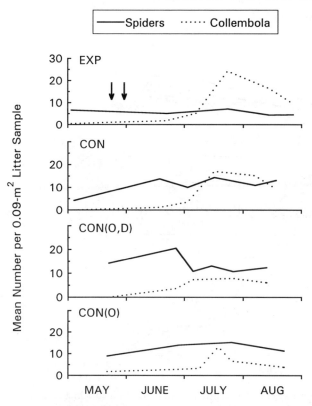

Fig. 6.2 Change in densities of spiders and Collembola in a fenced plot on the forest floor from which spiders had been removed (EXP), and in three different types of control plots: a 'complete' control (CON), which was a fenced plot in which the litter was disturbed as in EXP; a partial control [CON(O,D)], which was an unfenced plot in which the litter had been disturbed; and an undisturbed, open plot [CON(O)]. Spiders were removed at the end of May during the interval indicated by the two arrows. (After Clarke & Grant 1968.)

Collembola were the most numerically abundant prey. Although fluctuations in Collembola accounted for most of the variation in total numbers of potential prey in all four plots, Clarke & Grant did not apply statistics to determine whether Collembola densities in removal and control plots differed. The investigators concluded that such differences 'could not be tested in the same way as centipedes because the collembola, unlike the centipedes, were highly aggregated.' Variances of estimates of Collembola densities may have been so high that differences between removal and control plots are not statistically significant. This inconsistent application of ideally objective statistical criteria stems from

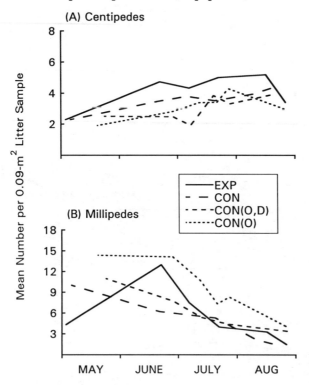

Fig. 6.3 Change in densities of (A) centipedes and (B) millipedes in a fenced plot from which spiders had been removed, and in three control plots (ref. Fig. 6.2). (After Clarke & Grant 1968.)

confusion over what constitutes the error variation against which variation due to treatment effects must be tested. In field manipulations such as Clarke & Grant's the experimental unit is the plot because it contains a single population that has received one treatment. Without replicating treatments of this unit, one cannot test hypotheses about treatment effects because the design provides no estimate of error variance.

Ecologists have usually appreciated the need for replication, but have often found it logistically difficult to replicate units of a size they judged to be sufficiently non-artificial. Clarke & Grant wrestled with this problem: '. . . given the time available for sampling (30 days), the need for large areas and a reasonably representative and frequent sampling in rain-free litter, replication was unfeasible.' Numerous ecologists have struggled to resolve this dilemma, but frequently with a clouded understanding of which hypotheses were actually being tested when standard ANOVA and χ^2 tests were applied to the results of field experiments. Hurlbert's (1984) major contribution to solving the problem has been to focus attention on what constitutes the experimental unit not only for the design of the experiment, but also for the associated statistical tests. Before Hurlbert clarified this issue, numerous field experiments were either inadequately replicated or samples within replicates were mistakenly defined as experimental units for the statistical analysis. Clarke & Grant, along with multitudes of fellow sinners among the ranks of experimental ecologists (this author included), should be forgiven for their transgressions into pseudoreplication; these were errors committed years before Hurlbert (1984) coined the term and brought the problem widely to the attention of field experimentalists. If Clarke & Grant had designed their study in the light of Hurlbert's admonitions against pseudoreplication, they would have reduced the size of their plots and sampled less frequently, which would have allowed them to replicate the experiment. With replication they would also have been able to determine whether or not spider densities really had been reduced substantially in the removal treatment.

The title of Clarke & Grant's (1968) publication suggests their experiment was a preliminary study. Unfortunately, additional experiments were never completed, perhaps because of the large investment of resources needed to conduct removal experiments on the required scale and with adequate replication. Despite its limited scope, their study introduced many ecologists (including this author) to the idea of experimenting with spider populations in nature. The experimental approach to the role of spiders in the forest-floor ecosystem that was pioneered by Clarke & Grant should be continued.

Results of field experiments in salt marshes (Vince, Valiela & Teal 1981, Döbel 1987), a grassland (Kajak & Jakubczyk 1975, 1976, 1977) and a Caribbean island (Pacala & Roughgarden 1984) provide good evidence that spiders limit insect numbers, though in these experiments spiders were not manipulated directly. These studies are discussed in detail with the treatment of indirect effects in Chapter 8. The experiment with *Schizocosa ocreata* (Wise & Wagner 1992; Chap. 5) indicates that the youngest instars of this lycosid may limit densities of Collembola, but more research is needed to determine the strength of the effect, and whether or not larger instars also limit their prey and suffer exploitative competition. Evidence collected so far is highly suggestive, but the hypothesis that spiders limit densities of insects deserves more intensive experimental investigation in natural communities.

Recent experiments with agroecosystems have demonstrated that spiders can be important in limiting densities of crop pests. These results are particularly intriguing, because spiders do not possess characteristics usually attributed to successful biocontrol species (Riechert & Lockley 1984). As generalist predators, spiders do not specialize upon particular pest species. Furthermore, spider fecundities are usually lower, and their generation times longer, than those of many insect pests. Riechert & Lockley (1984) consider it unlikely that spider species will closely track changes in prey populations; they suggest, however, that although individual spider species acting alone are not likely to be effective biocontrol agents, the entire complex of spider species in agroecosystems plays a major role in limiting the growth of pest populations.

Before examining the evidence for spiders as biocontrol agents, it would be helpful to review the components of predation that relate to the regulation of prey density, evaluate the evidence with respect to spiders, and briefly address the concepts of *population regulation* and *determination of population density*.

Regulation of population density

A predator has the potential to regulate a prey population only if the predator responds to increases in prey density by inflicting greater percentage mortality. Whether or not a population of predators causes such density–dependent mortality depends upon the nature of the functional and numerical responses, concepts introduced by Solomon (1949) and developed further by Holling (1959). The functional response is defined as the change in the rate at which an individual predator captures prey as prey density changes. The numerical response is the change in population density of predators as a function of changing prey density. Together these components of a predator's response to changes in prey density comprise the total response, which is expressed as a fraction of the prey population consumed. An increasing total response is a necessary but not sufficient condition for a predator to regulate a prey population, i.e. to maintain the prey population around an equilibrium density.

Holling (1959) defined three basic types of functional responses (Fig. 6.4). In the Type III, or sigmoid response, the rate of prey capture increases at an accelerating rate over a range of prey densities before effects of predator satiation or handling time cause the rate of prey capture to increase at a decreasing rate until a plateau is reached. This sigmoid-type response is the only functional response that can produce

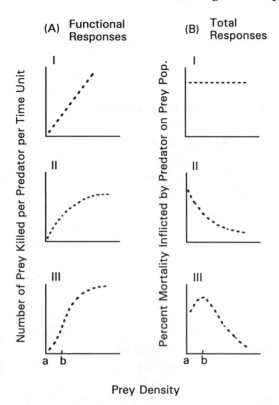

Fig. 6.4 (A) Three basic types of functional responses of a predator to changes in prey density. (B) The resulting 'total responses' if the predator displays no numerical response to increases in prey density. Only the Type III response can lead to regulation of prey density in the absence of a numerical response. Such a response is a necessary, but not sufficient, condition for a density-dependent total response, and leads to density-dependent mortality only at prey densities between 'a' and 'b'; beyond 'b' mortality is 'inversely density-dependent.' (After Holling 1959.)

density-dependent mortality in the absence of a numerical response (Holling 1959).

Functional responses exhibited by spiders

Riechert & Lockley (1984) predict that spiders, as sit-and-wait predators, would not be expected to exhibit Type III functional responses. Yet, after reviewing the literature, they conclude that several spiders display Type III responses. The papers reviewed by Riechert & Lockley actually provide stronger evidence for their prediction than they realized. Of the five papers that they cite, three provide no support for a Type III response by spiders. Riechert & Lockley cite Mansour, Rosen & Shulov (1980), who conclude that the functional response of the clubionid *Chiracanthium mildei* is 'an s-shaped or sigmoid curve,' yet the data of Mansour *et al.* are clearly Type II. Field studies by Kiritani & Kakiya (1975) and Dabrowska–Prot & Łuczak (1968b) are also cited as having uncovered Type III responses. Kiritani & Kakiya (1975) followed seasonal changes in densities of a lycosid and its planthopper and leafhopper prey in a rice paddy. No information directly relevant to calculating a functional response was obtained and their hypothesized functional response curves are Type II, not Type III. Dabrowska–Prot & Łuczak (1968b) conclude that the tetragnathid they studied preys preferentially on the more abundant mosquito species, which could lead to a Type III response. However, they have no direct indication that such a response did occur, and absence of a statistical analysis weakens their inferences. Because these last two studies were performed with natural populations, evidence for a Type III response would have been the most directly relevant to the issue of prey regulation. All other studies of functional responses in spiders, including those published after Riechert & Lockley's review, have been conducted in the laboratory.

Two of the studies cited by Riechert & Lockley (1984) did uncover clear Type III functional responses. Nakamura (1977) found both Type II and III responses by lycosids. The most pronounced Type III response occurred with leafhopper prey. The author speculates that accelerated rates of predation at higher densities resulted from greater prey activity due to increased interference among leafhoppers on the rice seedlings in the experimental container. Increased prey activity at higher densities also produced the Type III response observed by Haynes & Sisojevic (1966) for a crab spider. Hungry males displayed a Type III response, which was due to increased activity of the prey at higher densities and did not result from behavioral changes of the spider. Determinations of feeding rates at different prey densities were replicated, but variances are not given, so it is difficult to evaluate the fit of the data to the model. Females were not used because feeding patterns were judged to be too 'irregular.' However, because female spiders capture more prey than males, it is their behavior that is more relevant to the regulation of prey populations.

Behavioral variability characterizes many studies of spider functional responses. For example, Kajak (1978a) discovered a basic Type II curve for two orb-weaving species and a crab spider, though variability was so high that it is difficult to distinguish clearly between possible response models. High variability appears even when spiders have been starved for equal periods of time before being tested (Hardman & Turnbull 1980). In an earlier study of a lycosid feeding on vestigial-winged fruit flies, Hardman & Turnbull (1974) uncovered evidence for a weak Type III response, but not among all stages. Adult females displayed a Type II, with a hint of a Type III for some of the feeding durations employed in the experiment. Sub-adult females displayed a Type II with a decline in feeding rate at the highest prey density. Furuta (1977) also discovered a decline in feeding rate by an oxyopid at the highest density of third-instar larvae of the gypsy moth; this decline, found for both well-fed and starved spiders, appeared after a Type III response had been displayed at lower prey densities. Provencher & Coderre (1987) uncovered considerable variation in the response of an orb weaver (Tetragnathidae) and a clubionid to different densities of two aphid species. Polynomial regressions provided best fits to the data, and the researchers lament that '. . . variability in the data blurs the real shape of these curves.'

Type III responses in spiders appear to result more from increased availability of prey due to greater prey activity at elevated prey densities than from learning or modification in foraging behavior (e.g. Haynes & Sisojevic 1966, Nakamura 1977). A further example comes from Döbel's (1987) laboratory studies of feeding by two lycosids and a salticid on planthoppers abundant in salt marshes. His data fit models based on both Type II and III responses; improvement in fit is minor in most instances in which a Type III curve fits the data better. Feeding by the wolf spider *Pardosa* provides an instructive exception. Addition of *Spartina* thatch clearly shifted the response curve from a Type II to a Type III (Fig. 6.5). Döbel postulates that the thatch provides a refuge for a greater fraction of the prey population at lower densities, leading to a jump in feeding rate when prey numbers exceed a certain threshold.

A Type III response can also result from a predator ignoring rare prey and switching to alternative prey as they become more abundant (predator switching; Murdoch 1977). The sheet-web weaver *Linyphia triangularis* appears to reject some prey species the first time they strike the web (Turnbull 1960b). The agelenid *Agelenopsis aperta* also initially rejects novel prey (Riechert & Łuczak 1982). Such behavior appears to be independent of the relative abundance of other prey species and hence is not the type of switching that has the potential to produce a stabilizing Type III functional response. Studies of predator switching are rare for

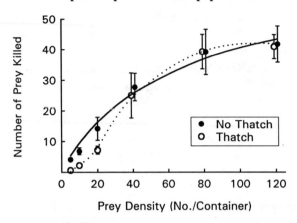

Fig. 6.5 Effect of a refuge from predation (thatch) upon the functional response of *Pardosa* (Lycosidae) feeding upon planthoppers in a laboratory experiment. Addition of thatch changes the functional response of the wolf spider from Type II to Type III. (After Döbel 1987.)

spiders. Provencher & Coderre (1987) found no evidence for switching in a laboratory study of spider predation on two aphid species; instead, relative activity of the aphid prey determined which species comprised the larger fraction of the spider's diet. Riechert & Lockley (1984) predict that although spiders can discriminate among prey of different profitabilities, switching should not be common because the temporal unpredictability of prey faced by most spiders should select against behaviors that cause them to reject any energetically profitable prey.

A Type III functional response induces density–dependent mortality in the prey population only at prey densities for which the curve accelerates; past the inflection point the percent mortality declines ('inverse density dependence'; Fig. 6.4). One cause of the declining rate of capture at higher prey densities is predator satiation. Riechert & Lockley (1984) point out two features of spider behavior that should retard the appearance of the plateau in the functional response: spiders digest their prey externally, and web builders can store captured but undigested prey in their webs. The behavior of killing more prey than can be utilized immediately could conceivably evolve in response to frequent bouts of prey scarcity of unpredictable duration, leading to caching of prey in the web for future use. Riechert (1974a) observed that given unusually high encounter rates with prey, *Agelenopsis aperta* accumulates 'more prey than they can plausibly consume before the prey desiccates and becomes difficult to ingest.' Riechert & Lockley (1984) refer to this type of behavior as 'wasteful killing' and suggest it may be common in spiders, citing seven additional studies as evidence. However, close examination of the studies in these papers does not reveal the degree of 'overkilling' that might be implied. For example, in their review Riechert & Lockley cite Kajak (1978a) as evidence that 'a spider will kill as much as 50 times the number of prey it actually consumes.' However, scrutiny of Kajak (1978a) suggests she did not discover such a phenomenon. Kajak found that an 800 × increase in the food supply was accompanied by a 50-fold increase in the rate of prey consumption, but she provided no data suggesting that almost all of these captured prey were never consumed, though it is likely that many usable calories in each captured prey individual were not extracted when prey were superabundant. One of the cited studies is that of Givens (1978), who reports that less energy is extracted from more recently killed prey when the spider has captured several prey in sequence. Riechert & Lockley seem to equate failure to extract all the energy from a prey item with wasteful killing, though this behavior is not necessarily wasteful, as the spiders are collecting usable calories from the captured prey.

Since the review of Riechert & Lockley (1984), truly wasteful killing of prey has been documented in at least one laboratory study of a spider (Smith & Wellington 1986). The orb weaver *Araneus diadematus* showed a Type II response over the lowest five prey densities employed, followed by a sudden increase in feeding rate at the highest density. The spider did not always consume some prey after killing and wrapping them. Hardman & Turnbull (1974) observed a similar jump in the rate of prey consumption at high prey densities by a male lycosid, though they presented no evidence that the rate of killing was wasteful. One wonders whether evidence of wasteful killing by spiders is a laboratory artifact. Such behavior probably does not characterize the functional response of most spiders within the range of prey densities encountered outside of the laboratory. In fact, the number of prey attacked and killed is more likely to decrease as the rate of prey supply is increased and the spider becomes satiated [pers. obs. during prey-supplementation field experiments (Wise 1975, 1979, 1981)].

It appears that spiders do not usually exhibit strong Type III responses to increases in prey densities. When present, Type III responses by spiders appear to result from changes in the vulnerability of prey to attack. Experimental results to date suggest that future research should elucidate the extent to which Type III responses by spiders occur in natural situations as a result of changes in prey behavior at higher prey densities.

Spider numerical responses

A predator has the potential to regulate its prey even in the absence of a Type III functional response if it exhibits a numerical response. Spiders show both aggregational and reproductive numerical responses to increases in prey densities in nature. Spiders have been shown to aggregate in habitats with higher prey densities, and temporal increases in prey density within a particular habitat can be correlated with increases in rates of spider reproduction. Examples of such responses, including findings of field experiments, are extensively discussed in Chapter 2 as evidence that spiders are food limited. It is possible that these numerical responses contribute to the regulation of insect densities. On the other hand, the numerical responses exhibited by spiders may not lead to long-term density-dependent mortality in prey populations. No total response curves [(functional response × numerical response)/prey density] have been published for spiders. Simply identifying numerical and Type III functional responses suggests, but by no means proves, that spiders contribute to the regulation of prey populations.

Do spiders regulate prey populations?

In most communities one randomly selected spider species would represent a relatively small fraction of the total spider biomass, particularly when the entire season is considered. Spiders are generalists that share prey not only with other spiders, but also with other predacious arthropods and vertebrates (e.g. Spiller & Schoener 1988). Thus removing a single species of spider should have minimal impact upon a particular prey population, much less upon the complex of prey species in the ecosystem. The only exception would be ecosystems where the diversity of both prey and spider species can be low, such as salt marshes (Vince *et al.* 1981, Döbel 1987).

As generalist predators with annual life cycles, spiders can be lumped conceptually into a predator complex when generalizing about their contribution to the dynamics of other components of the community. Spiders comprise a large fraction of the invertebrate predators in terrestrial ecosystems. Data already discussed show that, as a collective unit, spiders consume impressive fractions of insect populations in many ecosystems. Could this complex of generalist predators regulate prey populations? It is unlikely.

We should not expect the spider assemblage to regulate insect populations, i.e. maintain them around some equilibrium point. The entire collection of spiders in a community should not be any more effective than an individual species as an agent of density-dependent mortality. The absence of predator switching and the rarity of strong Type III functional responses among individual species should result in an entire complex that behaves no better than the best single spider species. Furthermore, numerical responses by spiders to increases in prey density may be inadequate for the entire spider complex to tightly regulate prey populations. Spiders exhibit aggregational responses to high-density patches of prey, but probably in most cases as a result of decreased emigration. *Agelenopsis aperta* can detect the airborne vibrations of flying insects at potential web sites before constructing the web (Riechert & Gillespie 1986), but it is unknown whether this ability is widespread among spiders, and, if so, whether spiders frequently locate high concentrations of prey before spinning a web. There is no evidence to date that most web spinners preferentially and rapidly immigrate into dense patches of prey. The arguments of Riechert & Lockley (1984) suggest that the high fecundities and rapid generation times of many insects will often enable them to escape control by spiders, because spiders will be able to inflict density-dependent mortality only over the lower range of prey densities, and only within one prey generation. By the time the spiders' reproductive responses result in more spiders, the prey population is likely to have increased even more, so that the percent mortality caused by spiders will not be higher. Hence the spider assemblage will be incapable of regulating insect populations.

Determination of population density

Doubts about prey regulation do not destroy the prediction that spiders limit insect densities, though they do force a certain clarification of concepts. Characteristics of the spider persona developed so far provide good reasons to predict that the spider has a major impact upon prey populations, if the distinction is made between *population regulation* and the *determination of population density*. We should not expect the spider complex to regulate insect populations around an equilibrium density, because spiders will only rarely be found to inflict strong density-dependent mortality upon their prey populations. The fraction killed is likely to be independent of density or may actually decrease with increasing prey density. However, because the spider complex does kill a substantial fraction of insect populations, insects should be more dense in the absence of spiders. Thus spider predation should be important in *determining* the density of insect populations, though spiders will not *regulate* insect numbers.

Density independence and density vagueness

Spider predation, as a density-independent biotic mortality factor, acts analogously to abiotic climatic factors in setting, or determining, the equilibrium density around which insect populations may be regulated by other factors that are density dependent, i.e. competition for resources, parasitism or disease (Fig. 6.6A). Strong (1984) reminds ecologists that this model of population regulation is too simple for most insect populations; it does not accurately reflect the relationship between density and population parameters in nature. Density dependence may only be significant at extreme densities; the variance in the relationships between density and population parameters may be so high that the pattern is density independent over the range of most insect population densities. Adopting Strong's approach, we should view spider predation as a major component of density-independent mortality in insect populations (Fig. 6.6B).

Spiders as biocontrol agents in agroecosystems

The behavior and life history characteristics of the spider persona led Riechert & Lockley (1984) to conclude that our leading character 'is a rather poor fit to the model pest control predator.' However, they argue strongly for the importance of the spider assemblage in limiting densities of insect pests in agroecosystems, postulating that spiders exhibit 'equilibrium point control' of insects, even though spiders will not closely track changes in pest densities. Several others who have reviewed the evidence agree that spiders are highly influential predators in agroecosystems. Nyffeler & Benz (1981) state that although periodic destruction of the vegetation in cultivated fields limits the potential of web builders as pest control agents, wandering spiders can reach high densities and 'may therefore stabilize certain insect populations of meadows and cereal fields.' Mansour, Richman & Whitcomb (1983) conclude that spiders can play a major role in limiting pest densities in cultivated fields and orchards. Young & Lockley (1985) argue that the lynx spider *Oxyopes salticus* has a major role in keeping several pest

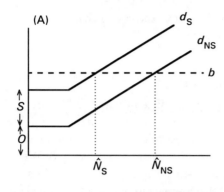

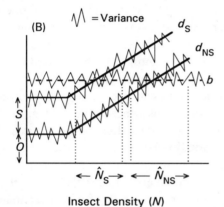

Insect Density (N)

Fig. 6.6 How the spider determines the abundance of insects in the ecological play. Predation by spiders, considered as a collective of generalist predators, causes substantial density-independent (DI) mortality in insect populations. The extent of this mortality is indicated by S; O represents DI mortality from all other factors. Regulation of insect densities results from the action of density-dependent (DD) factors, such as disease, parasitism or DD predation by predators that display stronger functional and numerical responses than spiders. In this model spiders are not the primary factors that prevent outbreaks in insect populations that could occur if conditions or resources changed in a direction that favored the growth of insect populations. However, a change in spider abundance, e.g. removal of the spiders from the system, causes an increase in average insect abundance (N_S to N_{NS}) by shifting the death-rate curve from d_S to d_{NS}. (A) A simple version of the model that ignores the variance associated with the relationships between insect density and birth/death rates. (After Enright 1976.) (B) A version of the model that better reflects the empirically determined relationships between density and population parameters. The variance results from the action of a multitude of factors and the effects of time lags. [After Strong's (1984) model of liberal (density-vague) population regulation.]

species at low levels, though they offer the caveat that this species 'alone cannot control major pest outbreaks.'

Indirect evidence: for and against

The abundance and diets of spiders in cultivated fields, orchards and forests suggest that spiders may contribute to maintaining low pest densities in many of these systems. Results of the precipitin test revealed that 5–20% of the spiders sampled by Loughton, Derry & West (1963) had fed upon the spruce budworm. Using ELISA techniques, Sunderland *et al.* (1987) determined that a wide range of predators feed on cereal aphids; the percentage of predators positive for aphids was high even at very low aphid densities early in the season. Predation by spiders was responsible for over 60% of the predation on aphids by the polyphagous predators that were sampled before the host plants had started to flower. Dean *et al.* (1987) found that the cotton leafhopper constituted over half of the prey items taken from spiders inhabiting woolly croton, the leafhopper's primary host. Numerous other surveys of agricultural systems and forests have shown that spiders often are abundant, and that pest species can form a substantial portion of spider diets [e.g. studies reviewed by Nyffeler 1982, Mansour *et al.* 1983, Riechert & Lockley 1984, Young & Lockley 1985, Nyffeler & Benz 1987 (the last paper reviews literature on temperate forests, not all of which are managed intensively enough to qualify as agroecosystems)].

Of course the appearance of pest species in the diets of spiders is not proof that spiders depress pest populations. In fact, some evidence suggests that spiders play only a minor role in pest suppression. For example, Nyffeler & Benz (1979b) found that total spider densities in cereal and rape fields were so low ($0.1–1/m^2$) that their impact on insect populations was 'almost insignificant.' In winter wheat and hay fields, micryphantid species dominate the spider fauna. They consume approximately 2% of the aphid population per day (Nyffeler & Benz 1988), which possibly represents a substantial mortality rate over the entire generation time of the aphids, but may not be as important as other mortality factors.

Furuta (1977) concluded that spiders in a pine plantation did not contribute either to the determination or regulation of density of macrolepidopterous pests: 'mortality caused by . . . (spider) . . . predation was negligibly small and inversely density-dependent.' Two oxyopids comprised a significant fraction of the spider fauna in the pine plantation

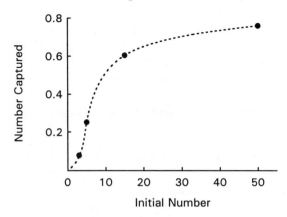

Fig. 6.7 Functional response of the oxyopid *Oxyopes sertatus* to changes in density of third-instar larvae of the gypsy moth, *Lymantria dispar*. Data from Table 6.2. No variance estimates are available because the number of spiders per experimental unit (= tree) was so small that samples were pooled in order to calculate capture rate. (After Furuta 1977.)

studied by Furuta. He conducted a field experiment in which larvae of the gypsy moth *Lymantria dispar* were added to pine trees with known numbers of *Oxyopes sertatus*. This oxyopid displayed a Type III response, but the average mortality of the gypsy moth was under 5% and was inversely density dependent over most of the range of larval densities tested (Fig. 6.7; Table 6.2). Bishop & Blood (1981) concluded that spider predation on cotton pests was not extensive enough to prevent crop damage in the system they investigated.

Spiders may consume so many other arthropod predators that the net effect on herbivorous pests may be minimal. In cotton fields of south Texas two lynx spiders (*Oxyopes salticus* and *O. viridans*) account for over 85% of the wandering spiders. These species feed upon cotton pests, but half of their diet consists of predacious arthropods, including other spiders (Nyffeler, Dean & Sterling 1987a,b). Spiller (1986b; Chap. 8) points out that interference competition between spider species may reduce the effectiveness of the spider complex in limiting pest numbers.

Trophic relationships between spiders and other arthropods are complex, and numerous factors affect insect population dynamics; thus, it is difficult to assess the contribution of spiders to pest control without manipulating the system. Because insecticides often affect spider populations, perturbations of spider populations have often been done unintentionally. Though not controlled experiments, such manipulations sup-

Table 6.2. *Mortality of third-instar larvae of the gypsy moth,* Lymantria dispar, *in a field experiment conducted in a pine plantation. Total mortality was high and strongly density dependent, but mortality caused by an abundant oxyopid,* Oxyopes sertatus, *was minor and was inversely density dependent*

No. of trees	Initial no. gypsy moth larvae/tree	Mean no. oxyopids/ tree	Total no. larvae captured by oxyopids	Mortality caused by oxyopids (%)	Total no. larvae died	Total mortality (%)
5	50	0.8	3	1.2	240	96
4	15	1.3	3	5.0	36	60
5	5	0.8	1	4.0	15	60
10	3	1.1	1	3.3	13	43

Source: Furuta 1977.

port the hypothesis that spiders influence pest densities. Good examples come from studies of rice paddies. The wolf spider *Lycosa pseudoannulata* is the dominant spider in rice fields studied by Kiritani *et al.* (1972). The diet of this lycosid consists primarily of two pests of rice, the green rice leafhopper and the brown planthopper. Estimated rates of predation by *L. pseudoannulata* upon the green rice leafhopper ranged from a few percent to 100%, depending upon the techniques used to calculate loss to the leafhopper population. Estimates of loss from all the spiders in the system ranged from *c.* 30% to 100% at low prey densities. Though the mortality among leafhoppers from spider predation was either density independent or inversely density dependent, the high mortality rates make it reasonable to predict that wolf spiders depress leafhopper populations. Kiritani & Kakiya (1975) cite results of Sasaba & Kiritani (1972) and Sasaba, Kiritani & Urabe (1973) that agree with this prediction. When pesticides were no longer applied to the rice paddies, lycosid densities increased and the density of eggs of the green rice leafhopper declined to half the original value. Similar evidence comes from a study in Philippine rice fields, though the experiment was not replicated. Kenmore *et al.* (1984) sprayed one rice field with insecticide and left another field, 500 m away, as a control. Densities of the rice brown planthopper were 800 × higher on the sprayed field; densities of spiders and veliid bugs, also predators of the planthopper pest, were lower in the insecticide-treated field.

Thus indirect evidence suggests that spiders contribute significantly to pest suppression in agroecosystems. Fortunately, several ecologists have employed field experimentation to test this hypothesis.

Field experiments

The pattern emerging from the results of field experiments suggests that spiders may indeed be important in determining densities of agricultural pests. However, field experiments with spiders as biocontrol agents have not always produced conclusive evidence because inadequate controls or lack of replication sometimes limit interpretations. Thus a review of experiments conducted to date will help place the evidence in perspective.

Kayashima (1961) reported on an experiment in which, over a four-year period, 45 000 *Oxyopes sertatus* were released in a forest plot in an attempt to control populations of a cecidomyiid pest. After four years the estimated tree damage was about 50% less than at the beginning of the release program, and about 75% less than in the control plot. The number of larvae that pupated in the control plot was four times the number in the plot to which the oxyopid had been added. Lack of replication and absence of information on initial numbers of larvae in the two plots severely weaken inferences about the actual impact of the introduced spiders.

Jones (1981) reports that the Chinese have used straw bundles as shelters for spiders to conserve their numbers during irrigation of rice paddies. This approach to spider conservation was associated with a 50–60% decline in pesticide use in 1977 over a 3000 ha region of Hunan Province, People's Republic of China. Unfortunately, it is impossible to evaluate probable causal connections underlying this correlation without more detailed knowledge of the control program.

In a replicated field experiment, Kobayashi (1975) indirectly increased the density of spiders in a rice paddy by adding fruit flies to adjacent dikes. The numerical response by spiders to increased prey in the dikes (discussed in Chapter 2 with other evidence for food limitation) led to an increased rate of spider emigration to the paddies. Apparently as a consequence, numbers of leafhoppers and planthoppers declined in the paddies. The decrease in rice pests did not occur rapidly enough to prevent damage to the rice plants, and subsequent increase in hopper densities suggests that the suppression of their numbers may have been temporary. Kobayashi suggests that spiders could be made more effective in controlling hopper pests if the provision of a continuous supply of supplemental prey induced a permanent numerical response by the spiders.

Correlations between spider density and survival rates of the rice brown planthopper (BPH) have implicated spiders as important actors in the saga of this pest in Philippine rice fields (Kenmore *et al.* 1984). These investigators conducted a field experiment with caged rice plants to determine whether or not spiders might depress BPH populations. All arthropods were removed by hand or suction apparatus from plants in one hundred 50 cm × 50 cm × 1.5 m cages of nylon mesh. After introducing 100 first-instar BPH nymphs into each cage, 50 of the cages were opened at the bottom to allow access to predators. Four of these opened cages had sticky traps to measure emigration rates of BPH nymphs and adults, which appear to have been negligible (1.6/cage/day). After three weeks leafhopper densities were an order of magnitude higher in the closed cages. During this period numbers of spiders and another arthropod predator, a veliid bug, were lower in the closed cages; however, after the increase in BPH numbers, predator numbers in the closed cages increased to levels similar to those in the open-cage treatments. This increase seems too rapid to have been a reproductive response by spiders, and must have resulted from movement of small spiders into closed cages that presented only a partial barrier to migration. Low rates of immigration by predators into the closed cages occurred during the first three weeks, when BPH densities were falling in the closed cages. BPH numbers continued to increase in the closed cages after predator densities were high, suggesting that much of the difference between closed and open cages may have resulted from restricted emigration. This outcome is somewhat perplexing, since the cage appears not to have prevented immigration by predators, but did restrict emigration by BPH (as judged by the rate of capture of BPH at the four open cages with sticky traps). More complete dissection of the results would require knowledge of the identity and size classes of predators, which may have differed substantially between the two treatments.

In contrast to these partially decipherable results, Oraze & Grigarick (1989) obtained clear-cut experimental evidence that the wolf spider *Pardosa ramulosa* consumes substantial numbers of the aster leafhopper, a pest in California rice fields. Spider densities were manipulated in small plots (0.8 m²) enclosed by aluminum rings that projected above the water surface. Oraze & Grigarick located plots in weedy areas within the rice paddies. In the first year's experiment they manipulated spider densities indirectly by placing a band of sticky repellent on one side of the rings. When placed on the outside, the sticky band prevented immigration of spiders, but did not prevent them from leaving the plots ('spider diminishing' treatment). Rings with the sticky barrier on the inside became the 'spider enhancing' treatment. This clever technique successfully altered spider densities; the difference between treatments was three-fold within three days, and six-fold by the end of the experiment. Although ingenious in conception, this procedure unavoidably affects movements of highly mobile predators other than wolf spiders. Oraze & Grigarick countered this problem the next year by manipulating lycosid densities directly. They placed sticky barriers on both sides of all the rings and removed as many *P. ramulosa* from each plot as possible in preparation to adding spiders in order to establish a range of densities.

Altering densities of wolf spiders affected numbers of aster leafhopper in both experiments (Fig. 6.8). Oraze & Grigarick were unable to elevate spider density within the rings above that found in the open habitat (c. $40/m^2$), most likely because of cannibalism. Although these experiments were conducted on small plots, two features argue for generalizing from the results to the open, more natural situation: (1) The size and composition of the plots resulted in natural densities of spiders in both the partially open (first year) and completely closed (second year) plots. (2) In both years Oraze & Grigarick utilized a randomized complete block design (four and three blocks for first and second year, respectively) in order to disperse experimental and control plots throughout the rice fields.

Japanese researchers have demonstrated that spiders not only consumed larval tobacco cutworm on taro in small field cages, but also lowered cutworm survival by causing the pest to leave its host plant (Nakasuji, Yamanaka & Kiritani 1973, Yamanaka, Nakasuji & Kiritani 1973). Lack of replication in one set of experiments, and the use of very small, confined aggregations of the larval cutworm in another one of their studies make it difficult to assess the actual magnitude of the spider effect in a natural system, but the experiments demonstrated the potential for significant depression of cutworm numbers by spiders.

A field experiment in an Israeli cotton field suggests that spiders can reduce densities of the Egyptian cotton leafworm (Mansour 1987). All arthropods were removed from four cotton plants and all but the spiders were returned; four undisturbed plants served as controls. Egg masses of

(A) 1984 Experiment

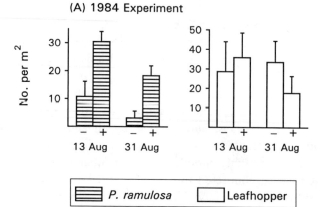

| ▤ *P. ramulosa* | ▢ Leafhopper |

(B) 1985 Experiment

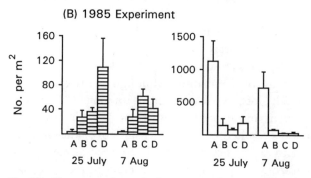

Fig. 6.8 Response of aster leafhopper numbers to manipulations of lycosid densities (*Pardosa ramulosa*) in 0.8 m² fenced plots. (A) 1984 Experiment: Lycosid densities altered indirectly by restricting immigration (−) or emigration (+) of spiders, starting 10 August. Spider densities were significantly different ($p < 0.05$) on both dates; differences in leafhopper numbers were significantly different only on the second date ($p < 0.01$). (B) 1985 Experiment: Lycosid densities altered directly by adding different numbers of spiders to cleared plots that had barriers to spider migration. Numbers added for A–D were 0, 22, 86 and 172, respectively. Adding lycosids significantly decreased leafhopper densities ($p < 0.05$ or 0.01), but leafhopper densities were not significantly different between any of the spider-addition treatments (Tukey's HSD method). (Based on data in Oraze & Grigarick 1989.)

the pest were introduced and the plants were enclosed in cloth bags. Five days later larval densities and leaf damage were higher on the plants from which spiders had been eliminated (Table 6.3). The pattern appears compelling, but it is not possible to determine whether or not the treatment differences are statistically significant because larval densities and indices of leaf damage from each plant have been pooled.

Table 6.3. *Effect of spiders upon survival of larvae of the Egyptian cotton leafworm on cotton plants. Larvae emerged from egg masses that had been attached to the foliage. All spiders were removed from four plants and were left on four others. Data on larval distributions and leaf damage have been pooled for the four experimental units within each treatment*

Treatment	Total no. leaves	Percentage damaged	Distribution of larvae on damaged leaves (%)[a]					Damage to leaves (%)[b]				
			0	1	2	3	4	0	1	2	3	4
Spiders present	233	23	10	65	25	0	0	0	82	8	7	3
Spiders removed	184	51	0	11	0	51	38	0	19	33	38	10

Notes:
[a] Code: 0 = no larvae present; 1 = 1–5 larvae/leaf; 2 = 5–10; 3 = 10–20; 4 = > 20.
[b] Code: 0 = no evidence of feeding on leaf; 4 = entire leaf surface nibbled.
Source: Mansour 1987.

Mansour and his colleagues had concluded from earlier field experiments that spiders are important biocontrol agents in orchards in Israel. In one study spiders and other arthropods were removed from three branches on a scale-infested citrus tree, and all but the spiders were returned (Mansour & Whitcomb 1986). Branches on a neighboring tree also infested with the scale insect were left undisturbed. The group of branches on each tree was enclosed with a cloth bag, and after two weeks the number of scale insects on the branches without spiders was over 5 × higher than on the undisturbed tree. Though the apparent impact of spiders was dramatic, the design of the experiment presents two potential problems: (1) treatments and controls were each confined to a single tree; and (2) the use of a closed system may have artificially increased the predatory impact of spiders by limiting pest emigration, and by excluding natural enemies of spiders, which could have substantially decreased their numbers in an open, less artificial situation.

In an earlier experiment Mansour *et al.* (1980b) established that removing spiders from apple trees increased the number of larvae of a leafworm pest, with resulting increases in leaf damage. The experimental protocol differed from that of Mansour & Whitcomb's experiment with the citrus pest. The earlier experiment was replicated across trees (three control trees, and three trees from which spiders had been removed and all other arthropods returned). Known numbers of egg masses of the insect pest were introduced to the trees. Unlike the study

Table 6.4. *Condition of egg masses of a pest of apple trees (*Spodoptera littoralis*), abundance of hatched larvae, and damage to infested leaf in response to manipulation of spider numbers. Each experiment lasted five days*

Treatment	Exp.	No. egg masses added[a]	Damage to egg mass[b]			No. larvae on leaf[c]			Damage to leaf[d]		
			0	1	2	0	1	2	0	1	2
Spiders present	I	20	1	1	18	20	0	0	17	3	0
	II	27	0	0	27	27	0	0	20	7	0
Spiders removed	I	20	14	0	6	6	10	4	6	8	6
	II	26	21	0	5	6	14	6	6	10	10

Notes:
[a] One egg mass added/leaf.
[b] 0 = no damage to egg mass; 2 = egg mass destroyed.
[c] 0 = no larvae present; 2 = all hatched larvae present.
[d] 0 = no feeding on leaf; 2 = entire leaf area nibbled.
Source: Mansour *et al.* 1980b.

with the citrus pest, Mansour and his colleagues did not enclose branches in order to prevent immigration of spiders into the removal treatments; instead, bands of sticky compound at the base of the spider-removal trees impeded recolonization by spiders. Bands of sticky substance were not applied to the control trees; hence, the possibility exists that improved prey survival on the spider-removal trees could have been caused by the exclusion of other predators. As in the experiment with citrus trees, there was no control for the initial manipulation of spider densities, i.e. arthropods, including spiders, were not removed and returned to the control trees. These problems in design are unlikely to have confounded interpretation of the results, because the experiment lasted only five days, and the egg masses of the pest were introduced after spiders had been removed from the trees. Mansour *et al.* (1980b) pooled the results from trees in presenting their results, making a statistical analysis unfeasible. Fortunately, they repeated the entire experiment a week later. Removing spiders clearly enhanced numbers of the pest, resulting in increased damage to the trees (Table 6.4).

The clubionid *Chiracanthium mildei* was the most common spider in the cotton and orchard ecosystems studied by Mansour and his colleagues. This species preys upon a wide spectrum of insect pests. *C. mildei* was the only spider regularly found near attacked mines of a leafminer pest in greenhouse experiments conducted by Corrigan & Bennett (1987). They suggest that *C. mildei* can detect the cryptic leafminer and attacks it by biting through the bottom surface of the mine. *C. mildei* may violate the conventional wisdom that single species of spiders cannot be effective biocontrol agents: it is a generalist predator that hunts a variety of insect species so effectively that it substantially reduces their numbers. It may be the major contributor to the suppression of pests by the entire spider complex in several agroecosystems, though this last hypothesis has not yet been tested by directly manipulating *C. mildei* populations. *C. mildei*'s role in pest suppression may result primarily from its abundance and active wandering in search of prey, since it does not display a Type III functional response in laboratory experiments (Mansour *et al.* 1980a).

The surprising effectiveness of single spider species in many experiments with agroecosystems may be an artifact of experimental design. Many of these studies employed completely caged experimental units, which perhaps artificially elevated the encounter rate between spider and prey. And most of these studies, even those with partially open plots or cages, were of short duration. Experiments conducted over an entire growing season might reveal less of an impact of a single spider species upon a particular pest, because of the spider's inability to track prey populations in cases in which fecundities and generation times differ. Longer-term experiments are clearly needed.

Chiverton (1986) examined the impact of polyphagous predators – beetles and spiders – upon densities of an aphid pest of spring barley in Sweden. Barriers (60 cm high) were erected around two plots early in the establishment phase of the pest in order to prevent immigration of ground predators from surrounding areas. Two open plots served as controls. Predator activity within all plots was monitored with pitfall traps containing water and detergent; use of traps also enabled Chiverton to remove beetles and spiders that already were within the fenced areas or that eclosed during the season. He constructed barriers around two additional pairs of plots at two later dates. Barriers constructed later in the season had less of an impact upon predator numbers because many had already immigrated into the field. The experiment was first done in 1981 with 5 × 2.5 m plots, and was then repeated in the two following years with 5 × 10 m plots in a different field. The most abundant predators were carabid beetles and spiders (mainly linyphiids).

Erecting barriers early in June had an effect on peak aphid densities; enclosing plots later in the season had no discernible impact (Fig. 6.9).

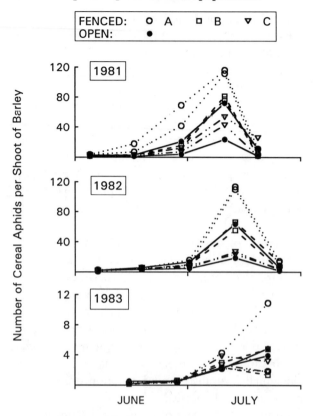

Fig. 6.9 Abundance of a cereal aphid (average number per shoot of barley) in plots from which beetles and spiders had been excluded by fencing, and unfenced controls. Barriers were erected around fenced plots at three different times: (A) early June, (B) mid-June, and (C) late June or early July. The experiment was repeated in three successive years. Note that aphid densities were an order of magnitude lower the last year of the study. Differences in aphid densities were statistically significant between the open plots and the 'A' fenced plots in 1981 and 1982. (After Chiverton 1986.)

Chiverton presents no statistical analysis of the data, but for 1981 and 1982 the peaks in aphid abundance in the early-enclosed plots differ significantly [p of one-tailed t-test $= 0.05$; approximate t-value calculated because variances are heterogeneous; statistic was calculated from values estimated from the figure in Chiverton (1986)]. All other comparisons are not statistically significant. Areas under the aphid-abundance curves do not differ significantly between enclosed and control plots for any comparisons. This pattern alone certainly is not overwhelming evidence, but Chiverton presents additional information that argues for the influence of generalist predators in this system. Significant negative correlations exist between peak densities of the aphid and cumulative numbers of predators trapped in 1981 and 1982 ($r = -0.87$ and -0.79, d.f. $= 6$). Spiders clearly contributed to this pattern; correlation coefficients between number of linyphiids trapped and aphid densities were -0.92 and -0.85 for 1981 and 1982, respectively. Effects of generalist predators were generally absent in 1983, when the aphid peak was an order of magnitude lower.

Two aspects of this study make it difficult to judge the extent of the release of the aphid population from polyphagous predators. The first is the variability between controls in 1981 and 1982, and between experimental plots in 1983. The second problem is the lack of a control for the effect of trapping. Because live traps were not used, trapping in the open plots reduced predator abundance to some extent. Possibly the impact of predation on aphids by spiders and beetles is greater than the results indicate.

Recent studies of Riechert & Bishop (1990) provide the best experimental evidence for the importance of the spider assemblage in agroecosystems. Their experiments demonstrate clearly that spiders can limit pest numbers in a mixed-vegetable garden system. Riechert & Bishop established four plots, each 70 m² in area, that were arranged in a row and were separated by strips of bare ground and metal flashing in order to impede movement of predators between plots. In order to encourage high numbers of immigrating spiders, no barriers were erected between the plots and an adjacent old field. Two plots were controls, with bare ground between rows of several types of vegetables. The other plots, the experimentals, were modified by placing mulch between all the rows, and by alternating rows of vegetables with rows of flowering buckwheat. The intent of both habitat modifications was to increase spider densities – mulch, by moderating physical conditions, and flowers, by attracting additional prey species. In the plots with mulch and flowers, spider densities were substantially higher, densities of insect pests were lower, and the extent of crop damage was less (Fig. 6.10).

This experiment has several positive features. Plots were large enough to minimize edge effects, and they were open, allowing migration across the boundaries not only of pests and spiders, but also other natural enemies that potentially affect interactions between spiders and their prey. Treatments were replicated, and controls and experimentals were randomly assigned with the constraint that they be systematically

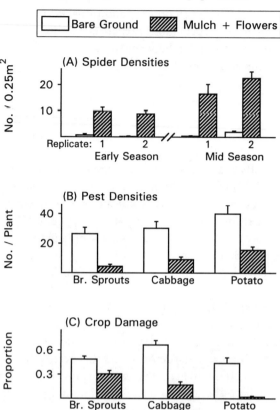

Fig. 6.10 Effects of adding mulch and flowers to a mixed-vegetable garden system. Source of spiders was immigration from an adjacent old field. (A) Spider densities in each replicate 70 m² plot, with standard errors based upon six samples from each plot. (B) Pest densities on those crops that suffered significant damage from pests. Standard errors based upon six randomly selected plants per plot. (C) Extent of crop damage from insect pests on plants sampled for pest densities. Mean proportion per plant of leaves or branches with > 30% damage. ANOVA's for all comparisons were performed on plot means, not sub-sample values (Table 6.5). [After Riechert & Bishop 1990; (B) and (C) give more detail than appeared in the publication; additional information on plot means supplied by Susan Riechert.]

interspersed. With only two replicates per treatment, conscious deviation from purely random assignment of experimental and control treatments helps avoid the 'nondemonic intrusion' of undetected spatial variation that could result from controls and experimental plots being segregated by chance (Hurlbert 1984). Plots were sub-sampled in order to improve estimates of spider density, insect numbers and plant

Table 6.5. *Results of analysis of variance of a field experiment in which spider numbers were altered in a mixed-vegetable garden system (Fig. 6.10)*

(a) Spider density – One-way ANOVA

	Source	d.f.	Mean square	F	p
Spring sample	Treatment	1	79.507	254.42	0.004
	Error	2	0.313		
Summer sample	Treatment	1	339.174	33.25	0.028
	Error	2	10.202		

(b) Plant damage – Split-plot design

	Source	d.f.	Mean square	F	p
	Treatment	1	0.670	313.50	0.003
	Replicate (treatment)	2	0.004		
	Vegetable	2	0.236	2.16	0.321
	Treatment × vegetable	2	0.109		

(c) Insect numbers – Split-plot design

	Source	d.f.	Mean square	F	p
	Treatment	1	1518.750	12.24	0.073
	Replicate (treatment)	2	124.034		
	Vegetable	2	160.433	44.17	0.032
	Treatment × vegetable	2	3.632		

Source: Riechert & Bishop 1990.

damage, but the variation between plots, within treatments, was used to calculate error variance. Spider density was estimated from six randomly placed 0.25 m² quadrats within each plot; differences in mean densities per plot were analyzed with simple one-way ANOVA (Table 6.5). Insect numbers and leaf damage were estimated on six randomly selected plants of each type of vegetable, making a split-plot ANOVA the appropriate statistical model for evaluating treatment effects (Table 6.5). In both analyses error mean squares have two degrees of freedom, which reflects the fact that experimental and control treatments each had only two replicates.

It should be heartening to empiricists that such dramatic responses can be observed and judged statistically significant with a design that relies on the minimum number of replicates possible. Only one F value, that for numbers of grazing insects, was judged marginal ($p = 0.073$). Not only is this value close to the conventionally accepted barrier of 0.05, the probability of this particular F, in the context of their experiment, is probably lower. It seems reasonable to employ a 'one-tailed' criterion, since Riechert & Bishop were entertaining only one alternative hypothesis – i.e. increasing spiders should cause insect numbers to decrease, not increase. As emphasized in previous chapters, one must analyze the total pattern of the results for consistency in evaluating evidence for or against a postulated phenomenon; we must avoid becoming hung up on either side of the 0.05 fence. Riechert & Bishop uncovered clear evidence that manipulating the habitat affected spider numbers, pest abundance and damage to the vegetation. They obtained their results with an experimental design that was most economical in number of replicates, and that utilized biologically realistic plots.

Despite the clear pattern in the outcome of the experiment, a potentially serious problem in interpretation remained. Spider densities were not manipulated directly. Could the habitat modification have caused the changes in insect numbers and plant damage by altering factors other than the spider assemblage? Riechert & Bishop expanded the design the following year in order to establish that the lower pest numbers indeed resulted from increased colonization of the plots by spiders and were not caused by changes in other factors (i.e. increases in other generalist predators or changes in nutrient availability) caused by mulching and the planting of flowers. They added a (mulch + flowers) treatment from which spiders were removed by hand. They were able to reduce average spider density by 50%, which was accompanied by an increase in pest numbers and the amount of insect damage to the vegetables. In the second year's experiment Riechert & Bishop also added separate treatments of mulch and flowers only, which revealed that it was the addition of mulch, and not flowers, that enhanced spider densities. Modifying the environment increased the densities of generalist predators other than spiders, but spider densities were over an order of magnitude higher. Also, the pattern exhibited by the other generalist predators did not match the changes in plant damage and pest numbers caused by the experimental treatments. Both of these findings indicate that most of the reduction in pest numbers and plant damage associated with the addition of mulch resulted directly from increased numbers of spiders.

A partially spun story

Field experiments support the hypothesis that spiders play a role in limiting insect densities. However, despite the results of several experimental studies, the detailed structure of this region of the spider's ecological web remains obscure. How extensive and how tightly spun are the connections between spiders and their prey populations? We still do not know the degree to which spiders limit insect densities in most terrestrial ecosystems. In natural (i.e. non-agricultural) systems, no long-term, well-controlled and replicated experiments have been conducted in which spider densities have been manipulated directly. Somewhat ironically, the best direct experimental evidence for the importance of spiders in limiting insect populations comes from studies of agroecosystems. Much of the experimental evidence in these systems, however, is based upon studies of limited scope – of short duration and with small, completely caged volumes of habitat. Such studies are valuable, and this approach of using field microcosms should not be abandoned. Nevertheless, many more experiments of the scope and approach of Riechert & Bishop (1990) are needed.

In order to understand the connection between spiders and their prey, we also must expand our experimentation to include other components of the ecological web. Schoener (1989) argues that in the Bahamian food web he and Spiller have studied, spiders are 'dead end' elements: they are affected strongly by lizards through predation and exploitative competition for prey, but spiders themselves have no strong effects on other elements of the system. Schoener points out that this particular food web may not be a good model for webs in other types of terrestrial communities. More research on this question is needed and would yield important results. Experiments with salt marshes (Vince et al. 1981, Döbel 1987; discussed in Chap. 8) highlight the complexity of interactions between spiders and different prey populations even in a supposedly simple ecosystem. Add other generalist predators and the web becomes distressingly complicated.

A more complete understanding of the extent to which spiders impact insect populations, and the underlying mechanisms, will elude us until we can analyze the results of numerous detailed studies, including field experiments, that focus simultaneously on as many trophic interactions as possible. We need to experiment with big pieces of the ecological web. In Chapter 8 I will explore this topic, but first I will detour briefly in order to address the question of how the physical environment influences the web of trophic connections we have been attempting to untangle. How is the spider's ecological web anchored?

Synopsis

Studies of energy flow and the relative abundance of spiders among top arthropod predators lead to the prediction that spiders substantially impact insect populations. Spiders capture a high fraction of herbivore net production in grassland systems. Spiders of the forest floor can consume over twice the mean annual standing crop of their prey. Studies of spider abundance and diets in agroecosystems suggest that spiders contribute to the limitation of insect pests in field crops and orchards.

A predator has the potential to *regulate* densities of its prey only if the mortality rate it inflicts is density dependent, which can occur if the predator displays a Type III functional response and/or a numerical response. Strong Type III responses are probably not common among spiders. Results to date suggest that most Type III responses by spiders result from elevated prey activity at higher prey densities rather than from learning or modification in foraging behavior by the spider. Thus the foraging behavior of spiders does not lead to the prediction that they are major regulators of insect abundance.

Several examples of numerical responses by spider populations to increases in prey abundance (discussed in Chap. 2), through either increased reproduction or an aggregative response, suggest that spiders could regulate prey populations. However, documentation that these numerical responses lead to significant density-dependent mortality in prey populations is lacking. Riechert & Lockley (1984) have argued that the longer generation times and lower fecundities of spiders compared to most of their prey, interference between spiders at high densities, and their euryphagous diet, all make it unlikely that spiders will closely track prey populations.

Apparent inconsistencies in generalizations about the role of spiders as predators can be reconciled if (1) population *regulation*, a density-dependent process, is distinguished from the potential role of density-independent mortality in *determining* population density, and if (2) the effects of a single spider species are distinguished from those resulting from the feeding behavior of all the spiders in a community. What is known about spider behavior, population ecology and the importance of spiders in food webs, leads to the hypothesis that spiders, *as a complex of generalist predators*, help limit insect populations by inflicting substantial density-independent mortality.

Riechert & Lockley proposed that the spider assemblage is important in suppressing pest populations in agroecosystems. The results of several field experiments support this hypothesis, but many more are needed because several of the studies were not replicated, not adequately controlled, or were of short duration with small, confined populations. However, the pattern emerging from the more complete field experiments suggests that spiders can limit insect pests. Manipulations in agroecosystems have implicated spiders as being important in a variety of field and orchard crops, though spider predation may not always be sufficient to prevent economically important damage. In some experiments investigators have augmented spider abundance by adding additional prey or by altering the physical structure of the habitat in order to increase rates of colonization by spiders. The other approach has been to reduce spider numbers on crop plants or trees. In several studies one or two closely related species of wandering spider – often a lycosid or clubionid – appear to be responsible for the majority of pest suppression attributable to spiders. The efficiency of a single species of spider in many of these experiments may reflect the short duration of the studies and the fact that the populations were confined in cages; hence, some of this research may provide an erroneous impression of the long-term effects of a particular spider. A recent long-term experiment with replicated open plots clearly establishes the importance of the entire spider assemblage in lowering pest densities in plantings of mixed vegetables.

Conventional wisdom originally dismissed spiders as not being good candidates for biocontrol agents, yet accumulating experimental evidence suggests that the spider community exerts significant pressure upon pest populations in some agroecosystems. On the other hand, ecologists have long suggested that spiders play a major role in the control of prey populations in more natural systems, yet widespread experimental evidence for this hypothesis is lacking. Some experiments, in which spider numbers were not directly manipulated by the investigator, provide strong indirect experimental evidence that spiders can limit prey numbers in non-agroecosystems (Chap. 8). A field experiment with a forest-floor community, in which spider numbers were directly reduced, is often cited as evidence that spiders limit their prey populations. This conclusion is unwarranted, however, because the treatments were not replicated.

Popular wisdom may be vindicated as the results of future studies accumulate: the spider complex may indeed contribute substantially to the limitation of insect populations in many natural terrestrial ecosystems. However, the hypothesis is by no means well substantiated. Many more carefully controlled and adequately replicated field experiments are needed.

7 · *Anchoring the ecological web*

Refining the metaphor: the web's non-trophic threads

Popular notions of the web of life emphasize feeding relationships between species. A famous example is Charles Darwin's speculation that the abundance of certain flowers might be related directly to the density of cats in the neighborhood. Cats eat mice, which prey upon the nests of humble bees, which alone bear responsibility for transferring pollen between flowers of certain plants. Darwin (1859) asserts '. . . it is quite credible that the presence of a feline animal in large numbers in a district might determine, through the intervention first of mice and then of bees, the frequency of certain flowers in that district!' The web may be even more intricate than Darwin imagined: cats eat birds, *which eat spiders*, which eat the bees that visit flowers in search of nourishment. Darwin must have inadvertently overlooked the spider; it certainly belongs in his 'web of complex relations' between 'plants and animals remote in the scale of nature'.

Contemporary community ecology continues to emphasize the contribution of biotic interactions to the structure of ecological communities, an emphasis that led naturally to the metaphor of spiders in ecological webs. In my mind, strands appeared most clearly as connections between spiders and their prey, and between spiders and their natural enemies. Because spiders are generalist carnivores they may simultaneously interact with each other as competitor, predator and prey. This maze of connecting threads resists easy untangling – a topic to be explored in the next chapter. Even if it is possible to resolve the complexity of indirect and cascading effects in the trophic web, another problem plagues the metaphor: connections with other components of the environment affect spider densities and distribution, and may substantially modify trophic relationships.

Without attachments to other points in ecological space, a metaphor woven solely of trophic threads would collapse into a formless jumble. The concept of an ecological web as a maze of trophic connections retains its form if non-trophic components of the environment are viewed as supports that anchor the metaphor. Their number and arrangement dictate the shape, complexity and resilience of the ecological web conceived as threads of biotic interactions. For example, abiotic factors such as temperature, humidity and wind may influence the type of habitat a spider selects for foraging, which constrains feeding options and helps to define its suite of natural enemies. Abiotic factors can affect mortality rates (examples discussed in Chapter 5), thereby influencing population density and the intensity of competitive interactions.

This chapter differs from earlier discussions of non-trophic factors by focusing upon the architecture of the spider's surroundings. Because some spiders spin traps to capture prey, parts of the metaphorical web can be anchored with the same structures used by spiders to support their real snares. Uetz (1991) points out that the physical structure of the environment also affects wandering spiders. For example, cursorial spiders with good eyesight rely upon mechanoreceptors to sense vibrations or to detect their surroundings through direct touch. The following discussion stresses *vegetation structure* and composition of the *leaf litter* as components of the architectural environment.

Correlative patterns

Vegetation structure

Ecologists have long recognized the close correspondence between the plant and animal species inhabiting an area. Early researchers realized that '. . . the physiognomy or physical structure of environments has an important influence on the habitat preferences of spider species, and ultimately on the composition of spider communities' (Uetz 1991).

Successional communities

A classic example is Lowrie's (1948) documentation of changes in the spider community associated with stages of plant succession on sand dunes along Lake Michigan. From the beach habitat Lowrie collected 17 species, predominantly wolf spiders that hide under debris during the heat of the day and hunt when temperatures are cooler. The most common was the lycosid *Arctosa littoralis,* which also was found in the foredunes, the next stage of succession formed by grasses that bind the sand and also provide attachment sites for webs of small and moderate-sized araneids and micryphantids (a family of small sheet-web weavers related to the Linyphiidae). A crab spider that builds its retreat among the grasses appeared frequently in the collections from the foredunes. The next plant-defined community back from the beach is the cotton-wood dune, which Lowrie apparently did not sample as extensively, but which he surmised has a spider fauna not much different to that of the foredune habitat. Farther back from the shore is the pine association, which shows an increase in number of spider species in all families. Most new species occur in the herbaceous layer. Lowrie collected slightly more than 40 species of spiders from the pine association. The black oak forest occurs next in the succession; this stage has the most undergrowth and the highest spider diversity (168 species) of all the successional stages sampled. Several families appear for the first time: linyphiids, oxyopids, pisaurids (close relatives of the lycosids) and mimetids, a family that specializes on eating other spiders. The beech–maple climax forest ranks second in diversity, with 122 species. Two uloborid species were among the 'most common and characteristic species' of this community; uloborids did not appear in collections from any other seral stage. Compared to the oak association, the beech–maple forest has fewer species, primarily due to its lower number of ground-dwelling spiders.

Lowrie (1948) speculates on what factors might contribute to the correlation between spider diversity and vegetation structure. He cites work by Shelford & Downing from early in the century showing that insect numbers increase from the beach habitat to sub-climax forest. It is unlikely that a greater diversity of prey contributes substantially to the coexistence of more spider species because: (1) results of field experiments suggest that exploitative competition for prey appears not to be common in forests and successional habitats (Chap. 4); and (2) as generalist feeders, spiders are more likely to be sensitive to changes in the physical structure of microhabitats than to differences in the diversity of available prey. Pirate spiders (Mimetidae) are a possible exception. Mimetids invade occupied spider webs and eat the owner, a specialized behavior that unavoidably depends upon the presence of web spiders. Pirate spiders were collected only in the oak and beech–maple forests. Lowrie believes that most successional changes in the spider community are responses to changes in physical factors, such as substratum (supports for webs and sites for retreats, etc.), temperature, exposure to wind, humidity and soil moisture.

Barnes (1953) collected spiders from a successional series of non-forest maritime communities along the mid-Atlantic coast of North America. Like Lowrie, Barnes discovered distinctive changes in the spider community that correspond with modification of the vegetation during

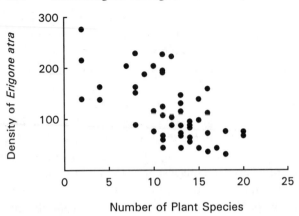

Fig. 7.1 Correlation between density of the micryphantid *Erigone atra* (mean no./plot) and the number of plant species during succession at Rothamsted. (After Duffey 1978.)

succession. Barnes also tested the consistency of the correlation between the species composition of the spider community and the nature of the vegetation by sampling 10 different sites representing the same succession-al stage, stands of the intertidal marsh grass *Spartina alterniflora*. He found the spider communities to be basically the same, likely due to the extreme homogeneity of the vegetation across sites. Does a similar consistency characterize examples of a successional stage that has more diverse vegetation? Barnes & Barnes (1955) investigated this question by sampling spiders in the herbaceous layer of 29 broomsedge (*Andropogon*)-dominated fields scattered over a 1000 km stretch of the piedmont of southeastern North America. The researchers collected over 5600 spiders from 85 species. The number of spiders in the sample for a particular site ranged from 99 to 345 (16 to 29 species). Barnes & Barnes did not calculate quantitative indices of similarity, but it is clear that, based on the number of species shared among fields and similarities in relative representation among different fields, the 'abstract broomsedge community' does have a corresponding abstract spider assemblage. Five species were judged to characterize this consistent association: a jumping spider, a lynx spider, a clubionid and two orb weavers.

Duffey (1978) suggests that plant species diversity influences density of the micryphantid *Erigone atra*. As evidence he cites its declining abundance throughout succession on the experimental fields at Rothamsted, England, a decline that is correlated with an increase in the number of plant species (Fig. 7.1). The spider's abundance clearly drops during plant succession, but increasing plant diversity may not have directly caused the decline in *E. atra*. Many factors potentially affecting web-building spiders change during plant succession.

Coulson & Butterfield (1986) studied a series of similar plant communities that did not form a successional series. They used multivariate techniques to characterize spiders caught in pitfall traps on 42 peat and grassland sites that ranged in elevation from 11 to 827 m in northern England. Species richness and diversity declined with altitude because fewer non-linyphiid species were found at higher elevations. Linyphiids comprised almost 75% of the total number of species trapped in their survey. Using cluster analysis to define spider associations, Coulson & Butterfield found that the sites separated into two groups at the level of 0.5 similarity. One group of spiders was characteristic of either mineral soils where grasses were the dominant vegetation, or of shallow peat soils where *Juncus* was dominant; the other grouping consisted of species from sites with peat soils where the dominant vegetation was *Eriophorum* and/or *Calluna*. At the 0.6 level of similarity the clustering procedure revealed two groupings of grassland and shallow peat sites, and six groupings of peat sites where *Eriophorum* and/or *Calluna* were dominant. Without these quantitative tools Coulson & Butterfield might have identified only two spider communities that could have been termed 'abstract' communities in the sense of Barnes & Barnes. Even these two associations displayed considerable overlap; Coulson & Butterfield conclude that '. . . very few common species of spiders were restricted to the grassland or the peat soils, and the most marked difference related to their relative abundance.' Nevertheless, the investigators recognize two major spider associations. They attribute the difference in composition of the two groupings to the height and architecture of the vegetation, and the water content of the soil.

Comparing distant communities: two approaches

Greenstone (1984) sampled the entire community of web-spinning species on sites distant from each other. He sampled meadow and scrub habitats at three elevations in Costa Rica and California by searching for spiders along 50 to 100 m transects and measuring the height of the vegetation every 2 m. Greenstone discovered a strong positive correlation between 'vegetation tip height diversity' and diversity of web spinning spiders (Fig. 7.2). As is true of such correlations, however, interpretation of underlying causes is not entirely straightforward. The difficulty lies in the possibly confounding influence of site differences not

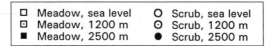

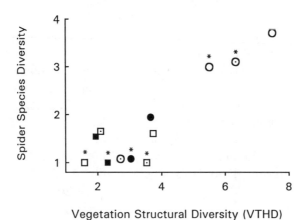

Fig. 7.2 Correlation between structural diversity of the vegetation ['vegetation tip height diversity' (VTHD)] and the species diversity of web-building spiders across a variety of habitats. Sites are from Costa Rica (*) and California. Vegetation heights were assigned to 20 cm classes in order to calculate a diversity measure based upon the Inverse Simpson Index, which also was used to define spider diversity. Least squares linear regression explains 83% of the variation in the data ($p < 0.001$). A similar relationship holds when VTHD is defined in terms of 40 cm height classes. (After Greenstone 1984.)

causally related to differences in vegetation structure. For example, the significant positive correlation comes primarily from the values for three scrub sites, two of which are at sea level. The third site is at 1200 m. Might a more moderate climate associated with lower elevations be causing the observed pattern? This interpretation is strengthened by the direct variation in spider diversity with elevation among Costa Rican scrub habitats (Fig. 7.2). Nevertheless, two aspects of the overall pattern argue against the importance of climate: (1) the mid-elevation scrub site in California has low diversity, and (2) among meadow habitats no relationship exists between spider diversity and latitude. There is also no correlation in meadows between spider diversity and plant structural diversity, largely because variation in the two variables is so low; in Costa Rica only one species of web-spinning spider was found in each of the three meadows. Spider diversity in California meadows is slightly higher, possibly because the California meadows are more protected than those in Costa Rica, which are heavily grazed (Greenstone 1984).

Despite these ambiguities, the consistency of several patterns in Greenstone's data provides good evidence for the contribution of vegetation structure to spider diversity. First, the strength of the regression comes primarily from the higher spider diversity of the scrub habitats. This pattern is reasonable – one would predict a greater variety of attachment sites in scrub vegetation. Considering only sea-level sites, one finds a clear positive correlation between spider diversity and structural diversity due to the high diversity of spiders in scrub habitats. Furthermore, there is a clear trend between spider diversity and vegetation structural diversity across all six scrub habitats (Fig. 7.2).

Greenstone attempted to remove effects of one possibly confounding variable, prey abundance; however, he found no correlation between insect abundance as measured by short-term sticky-trap catches and spider diversity ($p > 0.05$). A longer sampling period might have uncovered a significant correlation. Greenstone conjectures that a year-round measure of prey availability might be correlated with spider diversity, but he argues strongly for the major role of spatial heterogeneity in determining the diversity of web builders among the communities he sampled.

Rypstra (1986) uncovered significant effects of both vegetation amount and insect abundance upon the density of web spiders. She established plots in forests in three geographic regions: temperate eastern United States, sub-tropical Peru and tropical Gabon. Unlike Greenstone, Rypstra measured neither spider diversity nor structural diversity of the vegetation; instead, she determined the number of actively foraging web spiders and the total amount of vegetation in the herbaceous stratum. She also recorded the abundance of flying insects and the temperature at the time of sampling. A multiple regression analysis revealed that various combinations of these three variables accounted for the majority of the variation in density of active web spiders in all three habitats. If the analysis was restricted to either the midday or midnight period, $\geqslant 95\%$ of the variation could be explained by these variables. The amount of vegetation on the sample plots always was the most important variable, followed by insect activity. Rypstra argues that the relative importance of vegetation in her analysis may reflect the fact that data on vegetation was the most complete.

Rypstra sampled many (40 or 80) small plots within each of three forests, an approach that permits a clearer statistical analysis than

Greenstone's, in which correlations across widely separated sites possibly confuse variation in vegetation structure with variation in non-related variables. Greenstone's study, however, offers the advantage of having measured variables that directly reflect the diversity of spider species and plant architecture.

Perhaps it is not surprising that both studies uncovered a large effect of vegetation on spider populations, since one should find more web spiders wherever one finds more sites for web attachment. Failure to find a correlation would have been surprising, but only in the context of conventional natural history wisdom. Such wisdom is not always correct, and deserves the kind of quantitative tests that Greenstone and Rypstra applied.

What would we have concluded if they had found no correlation between spider diversity or density and vegetation characteristics? We might have concluded that factors such as prey abundance or natural enemies are more critical in explaining the structure of web-building spider communities, and that these factors vary independently of vegetation characteristics. We might also have concluded that web sites are superabundant. Both conclusions could have been wrong, just as we are not justified in concluding that the positive correlations uncovered by Greenstone and Rypstra indicate that web sites necessarily are a limited resource. A particular architectural configuration of the vegetation may be a necessary condition for an area to support certain spider species, but within an area of suitable vegetation the presence of more sites may not result in more spiders. Thus a positive correlation between vegetation structural diversity and spider species diversity may reflect the addition of new types of sites, without sites being limited in supply for each species.

Correlations between vegetation structure and composition of the spider community are not invariably tight. For example, Culin & Yeargan (1983) found that the spider community on soybeans, which was re-established every year when the new crop matured, often differed in species composition from one year to the next. This variation likely results from extensive disturbances in this non-equilibrium system, which would be expected to have a substantial impact upon the identities of foliage-dwelling spiders.

The several studies just discussed are examples of many that implicate vegetation structure as a factor determining the composition of spider communities. Lowrie (1948) mentions research that antedated his study. Uetz (1991) comprehensively reviews research in a variety of habitats that links composition of the plant and spider communities. Riechert & Gillespie (1986) survey numerous studies of habitat selection that implicate the influence of vegetation and associated environmental variables for particular species of spiders.

Factors correlated with vegetation structure

Changes in vegetation structure unquestionably correlate with changes in the spider community. The quantitative patterns just reviewed confirm and expand the intuition of many ecologists who have studied spiders. Anyone who has searched for a particular web spinner soon learns which vegetation most often supports populations of the sought-after species. For example, two common, widely distributed linyphiids, the filmy dome spider (*Neriene radiata*; Fig. 5.1) and the bowl-and-doily spider (*Frontinella pyramitela*; Fig. 1.5) frequently utilize openings in the foliage of junipers and similar evergreens. I have found that I can often locate the spider by first finding the appropriate shrubs. Others who have studied web spiders could relate similar examples. Plant architecture alone, however, cannot explain all the details of such patterns. For example: *F. pyramitela* occurs in wooded areas, but also is abundant on open, exposed vegetation. The filmy dome spider is apparently restricted to wooded or partially protected situations. Rarely are its webs found on exposed vegetation, such as on shrubs in northern bogs. Sometimes *F. pyramitela* appears to own such settings, where the webs of young instars present a glistening blanket of dew to early-morning naturalists. Is the web of *N. radiata* too filmy and fragile to withstand the wind, or does desiccation present a problem in exposed situations? Or have some subtle differences in architectural requirements for their webs escaped my attention?

The mechanisms by which vegetation affects habitat selection can be complicated. Several investigators have detailed this complexity by identifying the influence of associated variables. The distribution, growth and survival of foliage-dwelling spiders can be related to the influence of physical factors correlated with vegetation structure, i.e. problems of desiccation (Nørgaard 1951, Cherrett 1964, Horton & Wise 1983, Gillespie 1987), exposure of the web to wind (Eberhard 1971, Schoener & Toft 1983b), and exposure to insolation (Biere & Uetz 1981, Hodge 1987a,b). Diversity of spiders and representation by family differ along the elevational gradient of an intertidal salt marsh in parallel with changes in vegetation structure, amount of thatch and exposure to tidal flooding (Döbel, Denno & Coddington 1990). Factors associated with vegetation can interact to affect distribution on different scales, as the pattern for the orb weaver *Micrathena gracilis* demonstrates. Macrohabitat selection (deciduous over pine forest) and microhabitat selection within the deciduous forest reflect the degree to which gaps in the vegetation expose *M. gracilis* to direct sunlight and its web to the wind (Hodge 1987a,b).

Below I review in more detail selected examples of the different ways in which vegetation anchors the spider's ecological web.

A desert grassland spider

Susan Riechert and her colleagues have intensively analyzed the local distribution of the desert grassland spider *Agelenopsis aperta* (Agelenidae), uncovering several features of the vegetation and microphysiography that define this species' preferred web sites (Riechert, Reeder & Allen 1973, Riechert 1974a, 1976). In her initial studies Riechert used a technique developed by plant ecologists (block size analysis of variance) to uncover how the dispersion pattern changes with spatial scale. When the area of the sampling quadrat was small ($\leqslant 3$ m^2), the dispersion pattern was regular, consistent with the territorial behavior exhibited by this spider (Chap. 5). Evidence of aggregations at larger block sizes corresponded with several features of the habitat. Riechert (1976) confirmed her interpretation of the patterns exhibited by the block size analysis of variance by first measuring habitat features (vegetation and microphysiography) of both web sites and areas not used for building webs, and then employing multiple discriminant analysis to identify characteristics of sites where spiders constructed their funnel web. *A. aperta* web sites are associated strongly with shrubs, depressions in the ground, litter and flowering herbs. Shrubs are most important on lava beds; depressions, within which the spiders construct their sheet web, are an important defining feature of high-quality sites in mixed grassland. Shrubs and depressions are mainly critical in providing shade, not in providing structural support for the web. By modifying the thermal environment, vegetation and depressions allow the spider to forage longer, and hence capture more prey, than spiders on more exposed, lower-quality sites (Riechert & Tracy 1975). Litter also moderates the temperature for webs spun above it, because substrate covered by litter is cooler than lava or exposed soil.

In one study Riechert (1976) compared the lengths of time that spiders occupied sites of different discriminant scores. She marked 222 spiders, of

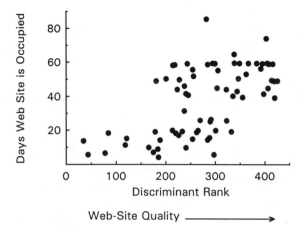

Fig. 7.3 Number of consecutive days a web site is occupied by the funnel-web spider *Agelenopsis aperta* in relation to the web-site quality as determined by a multiple discriminant score. The latter is assigned to a web site on the basis of how strongly its features match those that discriminate web from non-web sites in the entire study area. (After Riechert 1976.)

which 22 never moved, 93 disappeared (mortality or emigration), and 107 re-located their web site within the study area. Web sites of higher quality as measured by their discriminant score tended to be occupied longer (Fig. 7.3). Such a pattern could reflect differential mortality and not different rates of abandonment of sites of varying quality. Riechert was able to distinguish between these two possibilities because she had marked the spiders and thus could compare the qualities of sites abandoned with those of sites newly selected. She analyzed the moves of 65 spiders that she knew had moved more than twice, and discovered that they tended to move to higher-quality sites ($p < 0.01$; one-tailed sign test). This pattern suggests that the spiders, responding to some feature of the vegetation or microphysiography, recognize sites that are better than the one they abandoned.

Refuges on spruce branches
Vegetation structure also influences spider populations by ways other than constraining web construction or modifying the intensity of abiotic stresses. For example, by providing refuges the plant community might dampen interactions between spiders and their natural enemies. Gunnarsson (1988) discovered a difference in the spider communities inhabiting spruce branches that had lost different numbers of needles,

presumably due to the effects of air pollution. Branches with relatively more needles had more spiders over 2.5 mm long; Gunnarsson suggests that large spiders are more exposed to bird predation on branches with fewer needles. However, other variables could have caused the pattern. Gunnarsson compared trees from two different stands, one with relatively higher rates of needle loss than the other. The trees with fewer needles were in the more exposed stand, the one that was farther from the forest edge. Microclimatic differences or variation in predation intensity between the stands may have contributed to the higher abundance of larger spiders on branches with more needles.

A field experiment would provide the most direct test of the hypothesis that needle density affects the size distribution of spiders overwintering on spruce branches. Gunnarsson (1990) conducted an experimental test by selecting a pair of branches in each of 13 trees and removing 30% of the needles from one member of each pair. After two months the needle-thin branches had fewer spiders than the unaltered ones (66 ± 36 versus 102 ± 53, respectively; expressed as numbers/kg twig mass; $p < 0.05$, Wilcoxon matched-pairs signed-ranks test). Spider sizes, however, did not differ between treatments. Thus, Gunnarsson established the clear influence of needle density on spider populations, but did not confirm the original observation. He suggests that the short-term nature of the field experiment may have contributed to the failure to demonstrate a shift in size distribution of spiders.

Previous research (discussed in Chapter 5) documented the impact of bird predators on spiders inhabiting spruce branches, which gives support to the hypothesis that the architectural environment of the branch modifies the impact of natural enemies on spider populations. An expanded field experiment could directly test the postulated interaction between vegetation structure and bird predation. Altered and unaltered branches could be either enclosed with bird-netting or left open, with the experiment running over the entire winter. If bird predation is the direct cause of the needle-effect, survival on enclosed altered and enclosed unaltered branches should not differ, but would differ between treatments that are exposed to birds. Failure to find a statistically significant interaction term in the ANOVA would not support the bird-predation hypothesis, but instead would suggest that needle density affects survival by modifying behavioral interactions between spiders. Gunnarsson (1990) uncovered evidence for such an effect in a laboratory experiment, but modifying needle density in the laboratory setting did not influence spider numbers to the extent observed in the field experiment.

Vegetation may also affect trophic interactions along other threads of the spider's ecological web. Flowers can be critical resources because they attract insect prey. Chew (1961) suggests that a seasonal increase in the abundance of certain desert spiders that correlates with the onset of flowering results from an increase in prey. The presence of insect-attracting flowers is one defining characteristic of high-quality web sites for the desert agelenid *Agelenopsis aperta* (Riechert & Tracy 1975, Riechert 1976). Many species of thomisids wait in ambush on flowers. Douglass Morse has documented the importance of flowers for the crab spider *Misumena vatia*, which tends to forage on plants where the number of flowers or temporal pattern of flowering attracts the most prey per day (Morse 1981, 1986, Morse & Fritz 1982). When foraging on milkweed, *M. vatia* will leave a stem as its flowers senesce. Rates of prey capture on umbels of different quality can vary five-fold, the difference being directly attributable to differing rates of visitation by pollinating insects (Morse 1986).

The leaf-litter layer

Even after death, vegetation continues to tug at strands of the ecological web. Dead leaves and fallen stems potentially influence the distribution and abundance of surface-dwelling wandering and web-building spiders. Effects of the leaf litter on these spider populations have received particularly close attention.

Distribution differences between species can be correlated with preferences for different types of substrate, e.g. two lycosids of the genus *Pardosa* whose distributions in nature match laboratory-determined preferences for particular substrates (Hallander 1970b). Hagstrum (1970) reported that he captured the lycosid *Tarentula kochi* more frequently in pitfall traps located in areas of higher pine litter, although from the data he presents it appears that much of the pattern results from (1) the fact that more traps were located in areas where litter was present, and (2) immature spiders apparently avoided areas where litter was completely absent. It is reasonable to expect that the amount and structure of the litter will affect instars differently because of differing susceptibilities to desiccation and predation. Edgar (1971) suggests that immature *Lycosa (Pardosa) lugubris* avoid falling prey to larger instars by running inside the leaf litter, since adults run on top of the dead leaves. In a laboratory study Edgar (1969) found that adding artificial litter to a container retarded the rate at which large wolf spiders located and ate smaller individuals.

Properties of the litter account for differences in the courtship behavior of two similar lycosids of the genus *Schizocosa* that prefer microhabitats with different substrates. *S. rovneri* lives in floodplains where the leaf litter is more compressed than the deeper and more complex litter of upland areas, where the congener *S. ocreata* is more abundant (Uetz & Denterlein 1979, Cady 1984). Males of both species employ substrate-coupled stridulation to court females, and it appears that differences between the ability of the two litter types to conduct vibrations explain species differences in the reliance on visual and vibratory cues during courtship (Uetz 1991).

Broad correlations have been uncovered between the spider community and changes in litter cover during succession [e.g. early studies such as Lowrie (1948) and others reviewed by Uetz (1991)]. Some comparative investigations have focused specifically on the litter community. For example, Gasdorf & Goodnight (1963) found that spider densities and number of species were higher in the litter of a climax oak–hickory forest than in a floodplain forest. This study constitutes a preliminary investigation, because it was based upon only two 2500 cm² litter samples collected biweekly for 16 months.

Jocqué (1973) compared the spiders in the litter layer of a planted beech forest to those of a woodland with a higher tree diversity. Over 95% of the litter in the first forest was beech; decomposition was slow, resulting in a 7-cm-deep litter layer over a typical mor humus. In contrast, leaves of seven tree species accounted for 95% of the litter in the mixed woodland; decomposition was rapid, with a litter layer of 0–2 cm over a mull-like moder humus. The coefficient of similarity between the two spider faunas was 40%. Species diversity, measured either as species richness or equitability of relative abundances, was higher in the mixed woodland (respective paired values of S and H' for beech forest and mixed woodland were 55, 3.14; and 60, 4.11).

Communities of wandering spiders: three related studies
The most comprehensive studies of the relationship between characteristics of the leaf litter and the spider community are those of George Uetz and his colleagues, who have documented variation in communities of wandering spiders.

Two of their investigations examined changes in spider communities along a major environmental gradient. One study dealt with a gradient of flooding in a streamside deciduous forest in eastern North America (Uetz 1976, Uetz *et al.* 1979). Pitfall traps were arranged along an elevational gradient from floodplain forest (25% probability of flood-

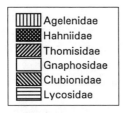

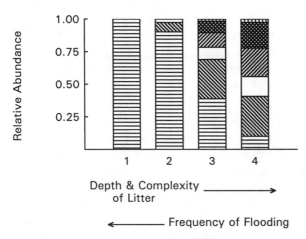

Fig. 7.4 Changing relative representation of spider families sampled by pitfall trapping at points along an elevational gradient. 1 = floodplain; 2 = transition zone; 3 = lower uplands; 4 = uplands. [Note: Although the Agelenidae and Hahniidae are web-building families, both sexes of several species are caught frequently enough in pitfall traps to suggest that they spend significant amounts of time foraging off the web (Breymeyer 1966, Uetz 1975)]. (After Uetz 1976.)

ing; forest dominated by silver maple) to an upland white oak forest that never flooded. Along the transect the forest floor varied from damp, silt-covered litter separated by bare patches to a continuous cover of leaves 3–5 cm deep. Lycosids dominated the floodplain, whereas upland areas had a much higher diversity of wandering spiders, with wolf spiders clearly a minority (Fig. 7.4). Flooding directly affects the spider community by destroying stationary egg sacs and forcing spiders to emigrate. Uetz points out that wolf spiders are better adapted to disturbed situations because they carry their egg sacs with them, and because they can rapidly recolonize an area after the effects of the disturbance have receded. Flooding indirectly affects spider densities by influencing the depth and structure of the leaf litter. Litter can provide refuges from predators, such as lycosids; ameliorate the physical environment; furnish crevices for the attachment of egg sacs; or furnish attachment sites for web construction (two of the families sampled by Uetz's traps, the Agelenidae and Hahniidae, build webs).

The second study relied on a less well-defined gradient (Bultman, Uetz & Brady 1982). They sampled the cursorial spiders (including vagrant species of the web-building agelenids and hahniids) along a gradient of three successional habitats: old field, oak forest, and beech–maple forest. The number of species was the same in the old field and oak forest (21), but about half as many in the climax beech–maple forest (11 species). Wolf spiders were most abundant in the old field, but were rare in the beech–maple forest. Although species richness was the same in field and oak forest, H' and J' (the evenness measure of species diversity) were slightly higher in the oak forest. The investigators cite this trend as being in agreement with Huhta's (1971) finding that the species diversity of wandering spiders increases during succession as depth of the litter increases (Bultman et al. 1982). Unfortunately for the trend, H' and J', along with species richness, were lowest in the climax forest. The low species diversity in the climax forest, which presumably has the deepest litter, results from low numbers of gnaphosid and lycosid species. The researchers attribute the decline in diversity to the manner in which litter structure affects foraging behavior of these two families, wolf spiders in particular: 'the development of a thick and intricate litter system during succession may prevent lycosids from dominating the cursorial spider community of climax forests' (Bultman et al. 1982). Although plausible, this hypothesis finds only weak support in their data, because so many other factors vary between meadow and forest that it would be risky to conclude that differences in the structure of the litter is the main cause for the pattern in species diversity. Similarly, the clear pattern observed by Uetz (1976) along the altitudinal gradient could have been caused by direct effects of physical disturbance rather than by differences in litter conditions. More direct support would come from correlations between litter characteristics and the species diversity of spiders in one type of plant community.

Uetz's first study, which preceded the two just discussed, uncovered such a relationship (Uetz 1975). He found a striking correlation between litter characteristics and spider diversity in a 14 ha tract of mature deciduous forest in which oaks, maples and tuliptree were common canopy trees. Over a summer's study, 34 species of wandering spiders in eight families appeared in a network of 15 pitfall traps, each of which had

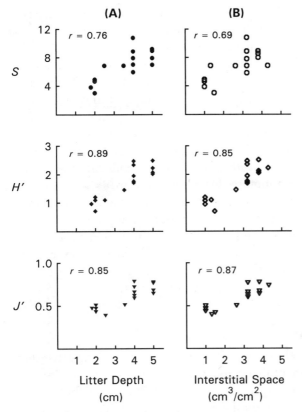

(A)   **(B)**

Fig. 7.5 Relationship between measures of species diversity of wandering spiders and (A) litter depth and (B) a measure of relative amount of interstitial space within the litter. S = species richness (total no. of species); H' = Shannon–Weaver Index, which incorporates S and relative abundance; J' = the evenness component of species diversity (H'_{obs}/H'_{max}). H' is maximum for a given S when relative abundances of all species are equal. All correlation coefficients are statistically significant ($p < 0.005$). (After Uetz 1975.)

been placed randomly within a 50×50 m grid. Any single trapping station caught a maximum of 12 species over the entire sampling period. Litter depth surrounding the 15 trapping stations ranged from 2 to 5 cm, and a measure of interstitial space within the litter (Uetz 1974) varied from 1 to 5 cm³/cm². Species diversity (H') was clearly higher in areas with greater litter depth and greater proportion of interstitial space, due to an increase in equitability of relative abundances as well as an increase in the number of species trapped (Fig. 7.5). Litter depth and Uetz's measure of interstitial space are themselves highly correlated ($r > 0.9$;

Uetz 1979), so that the observed correlations between spider diversity and these two aspects of the litter layer may reflect the same underlying cause. Complicating the interpretation further is the possibility that variation in litter depth is correlated with extraneous variables in a fashion similar to what may occur along the gradient from floodplain to upland forest analyzed by Uetz (1976). For example, in areas of the study site where the litter is not deep, it is shallow and compressed not only because of the leaf type, which is swamp chestnut oak, but also because water accumulates (Uetz 1979). Hence frequency of disturbance to the litter as well as litter depth and complexity may influence the composition of the spider community. Uetz points out that several environmental factors that may affect the activity-abundance of wandering spiders vary with the depth and structure of the litter: prey abundance, temperature fluctuations, moisture and refuges from predation. Because Uetz failed to find correlations between species diversity and temperature or moisture at the different trap locations, he concluded that the 'physical structure of the litter might be important.'

Field experiments

Abundances and diversity of spiders clearly correlate with variation in amount and structure of living and dead vegetation. In some situations causal connections are apparent, or at least appear so, but enough potentially important environmental factors vary with architecture of the habitat that experimentally manipulating them offers the best hope of identifying actual causes. Uetz (1991) summarizes the correlative studies as having 'concluded that spider species show highly specific associations with certain plant species and/or structural strata, and that where these habitat resources are more abundant, spiders are also. The basis of such associations is unclear, as few studies have considered or experimentally tested whether structure or other variables such as prey or microclimate were involved.' Several ecologists have experimentally manipulated features of both living vegetation and the leaf litter in an attempt to separate direct cause from indirect correlation.

The leaf-litter layer

The leaf-litter system presents the experimentalist with some of the greatest challenges in teasing out important factors, since plant detritus affords refuge for both hunted and hunter, profoundly affects the

Table 7.1. *Response of the spider fauna of grassland litter to different levels of experimentally induced trampling. Each value represents the mean of 25 litter bags exposed to trampling for one year. Each tread was equivalent to the step of a 82 kg person*

	Control	5 treads/month	10 treads/month
Litter depth (cm)[a]	4.2	?	1.6
Percentage open space[b]	63	54	38
Number of spiders[c]	275	74	42
Number of species	25	12	6

Notes:
[a] Number of litter bags upon which values are based is unknown. Author gives 1.6 cm as the depth of the trampled litter, but makes no distinction between the two treatments.
[b] Mean of scannings of gelatine-embedded sections of three litter bags from each treatment.
[c] Totals for 25 litter bags.
Source: Duffey 1975.

microclimate, and together with the associated microflora constitutes a source of food for the prey of spiders. A few undaunted researchers have manipulated the litter experimentally in order to uncover how the litter environment affects spider populations.

Duffey (1975) reports the results of two experiments. In one study, designed to examine the impact of trampling by humans upon the fauna of the litter, Duffey placed sterilized hay into 75 nylon bags, each $20 \times 30 \times 5$ cm with 1 cm openings. The bags, placed in a field in a randomized block design, received one of three treatments for a year: 5 treads per month, 10 treads per month and no trampling (control). Duffey did not present a statistical analysis of the data, but the pattern of the results suggests that the decrease in open space within the litter lowered the abundance and diversity of spiders (Table 7.1).

In another study Duffey (1975) investigated the effect of litter depth upon the density and diversity of woodland litter spiders. He found that in an oak–ash woodland, the 'multistem growth of old coppice stools' trapped falling leaves and woody debris, which reached a depth of 30 cm. The coppice stools had more spiders per equal volume of leaves than litter collected from the forest floor ($p < 0.05$, Mann–Whitney U-test). Duffey speculated that the difference might reflect the different composition of the debris in the coppice stool, not the greater depth of the

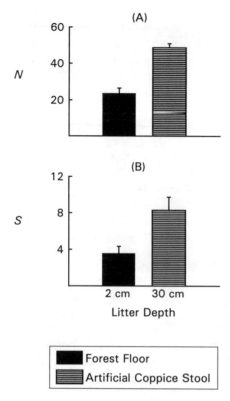

Fig. 7.6 Numbers (*N*) and species richness (*S*) of spiders in litter from artificial coppice stools (*n* = 4, litter depth = 30 cm) compared with values for equal volumes of litter collected from the nearby forest floor (*n* = 5, litter depth = 2 cm). Each sample = 18000 cm³ of litter; spiders were extracted in a heat-gradient apparatus. (A) Number of spiders per sample ± s.e. (B) Number of species per sample ± s.e. Differences between coppice stool and litter from the forest floor are statistically significant (*p* < 0.001). (From data in Duffey 1975.)

coppice-stool litter compared to the layer on the forest floor. As a direct test of this hypothesis, Duffey constructed four artificial coppice stools, filled them with fallen leaves, and left them on the forest floor for 15 months. A comparison with equal volumes of leaf litter collected from the ground nearby revealed a greater density of both individuals and different species in the deeper litter (Fig. 7.6). The apparent effect of litter depth is dramatic, but it is uncertain whether or not the difference reflects litter depth *per se*, or results from the fact that each coppice stool had 32 rods, each 30 cm high, which likely functioned as web supports. Similar structures were not available to spiders outside the coppice stools. A more direct test of the influence of litter depth would have been to house different amounts of litter in baskets of wide-mesh wire, including one treatment with normal amounts of litter to compare with an equal volume of entirely undisturbed leaves. The effects of the rods on the outcome of Duffey's experiment could be evaluated if it were possible to remove from the analysis all obvious web-building spiders that do not occur normally in the litter layer. One cannot refine the analysis in this manner with the present data because, although Duffey mentions some of the spiders that were restricted to one sample type, he does not present a complete list of species.

Litter manipulations on a large scale
The soil ecology research group of the Second Zoological Institute at the University of Göttingen has conducted a long-term litter-manipulation experiment in a German beech forest (Schaefer 1988). The study was designed to examine all components of the litter/soil system, not only spiders. The researchers manipulated litter depth in two large, fenced plots, each 10 × 10 m, and used a similarly-sized fenced plot as a control. Netting placed over one of the experimental plots each autumn excluded falling leaves. The other manipulated plot received additional litter; falling leaves collected by netting from nearby areas of the forest were added to this plot in order to make the rate of litter input 5 × normal. This treatment was imposed at the beginning of the experiment and was repeated the second year only. The control plot received normal leaf-fall each year.

The design, unfortunately, was not replicated. It would have been better to have used plots of 50 m² so that each treatment and the control would have had two replicates. Replication was apparently sacrificed in order to create plots that were large enough to minimize edge effects, and that enclosed realistic pieces of habitat which could be sampled intensively over many years. However, without replication it is not possible to assess statistically whether or not observed differences between plots reflect effects of litter depth. One can use knowledge of natural history to evaluate whether or not the results are what one might expect, but any statistical comparisons of plots in order to test the effect of litter depth would constitute pseudoreplication. Nevertheless, it is worthwhile to examine the results, because of the duration and sampling intensity of this experiment.

Average biomasses of spiders (mg dry weight/m²; web builders and vagrants combined) over a four-year period were 23, 116 and 208 mg/

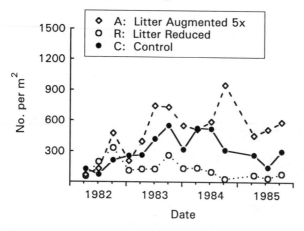

Fig. 7.7 Change in density of spiders on 100 m² plots on which litter fall was experimentally manipulated. Starting in 1981 litter was excluded from plot R, allowed to fall at natural rates in the control plot C, and increased 5 × in plot A (in 1981 and 1982 only). Spider abundances were estimated by extracting litter/soil samples using a Kempson apparatus. Nine families were represented, the most abundant being Micryphantidae, Linyphiidae and Theridiidae. (After Stippich 1989.)

m² for litter-reduction, control and litter-addition plots, respectively (Stippich 1989). Fig. 7.7 summarizes the change in density of individual spiders in each treatment over a four-year period. Stippich gives no data on species diversity, but she lists the relative importance of the families that were collected. Reducing the amount of litter eliminated the clubionids and amaurobiids, increased the relative representation of the micryphantids, and decreased the importance of the theridiids and linyphiids. Increasing the amount of litter did not have a large effect on relative representation of the different families.

Stippich sampled spiders by extracting them from litter/soil samples with a Kempson apparatus (Kempson, Lloyd & Ghelardi 1963, Schauermann 1982). This procedure may favor the collection of the less-mobile web spiders over wandering spiders, because the latter may escape when litter is being collected. In addition, most cursorial forms cover a larger area than that of a single litter sample. Stippich's samples contained no lycosids, very few salticids, and relatively few clubionids and agelenids. Differences between the spider communities of the German forest and those studied by Uetz and his colleagues could reflect major differences in species composition rather than different biases of the particular sampling procedures utilized in the studies. Using pitfall traps in conjunction with the Kempson procedure would have furnished valuable additional information on the spider community of the German beech forest. Most likely the decision was made not to use pitfall trapping because of the potentially negative impact on the entire research program. Using traps that killed the animals would have been too disruptive to Stippich's research and the additional ongoing studies of faunal interactions being conducted by other members of the research team.

It is striking that increasing the amount of litter 5 × appeared to have had relatively minimal effects upon spider abundances. Preventing normal litter accumulation appeared to have had a larger impact, but effects did not surface until a year after the litter manipulations began (Fig. 7.7). Removing litter in this experiment apparently did not substantially alter the abundance of Collembola (Wolters 1985), leading Stippich to suggest that experimentally reducing the amount of litter affected spider numbers by modifying architectural features of the litter and altering the microclimate.

Uetz (1979) also conducted a field experiment that tested the influence of litter depth on the spider community. His experiment focused on spiders, and was of shorter duration than the study conducted in the German forest, but also employed large plots. Uetz's experiment was well replicated. Uetz raked leaves in 10 × 10 m plots in order to create five different litter depths: 0 × , 0.5 × , 1.0 × , 1.5 × and 2.0 × the average depth of the litter in that part of the forest. He arranged 15 plots throughout a 50 × 100 m area in a randomized block design, with the provision that each experimental area be surrounded by non-disturbed 10 × 10 m plots. A pitfall trap in the center of each of the 15 experimental plots sampled the cursorial fauna in alternate weeks for four months. Litter levels were manipulated in March, when Uetz judged the leaves would not contain many wandering spiders or their egg sacs, though this was not tested. Uetz assumed that in adding litter or taking it away, he was not substantially altering spider densities because most overwintering spiders were probably under logs or somewhere else than the leaf litter. The open plots permitted unrestricted movement of spiders throughout the experiment, which lasted until October. Thus even if the litter in March housed significant numbers of spiders, it is reasonable to assume that the manipulation procedure itself did not cloud the results.

Experimentally modifying litter depth influenced the number of species of wandering spiders trapped (Fig. 7.8); the pattern was similar to that uncovered in his earlier non-manipulative studies (Uetz 1975, 1976).

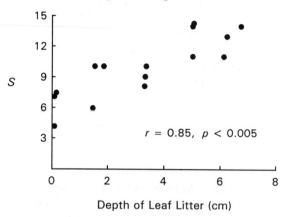

Fig. 7.8 Effect of experimentally altering depth of leaf litter upon the number of species of wandering spiders trapped. S = number of species per pitfall trap. (After Uetz 1979.)

This similarity suggests that in the two purely correlative studies, species diversity of spiders was causally connected with litter depth, not coincidentally correlated with a functionally unrelated variable. Manipulating the litter did not influence the total number of individual spiders trapped, an intriguing result which suggests that the increased species diversity was not the consequence of a simple area/diversity relationship; i.e. species diversity was not higher in the plots with more litter just because spider activity/abundance was higher here, so that more rare species were represented. Instead, some aspect of the increased amount of litter influenced the spider community by either affecting interactions between species and/or by expanding possibilities for individual species because levels of limiting factors had been altered. This lack of an effect on total numbers of spiders contrasts with Stippich's (1989) findings, but the extremes of litter depth were greater in her study.

Although Uetz's field experiment proves that the amount of litter itself affects species diversity of wandering spiders, mechanisms remain obscure unless the influence of factors correlated with litter depth can be separated. Uetz attempted this separation by calculating partial correlation coefficients between spider diversity and several independent variables: litter complexity, energy content of the litter, prey diversity, temperature variation and moisture variation. These parameters were measured three times during the experiment, and partial correlation coefficients were calculated for the seasonal totals (Fig. 7.9). Complexity of litter structure, followed by diversity of prey, influenced spider

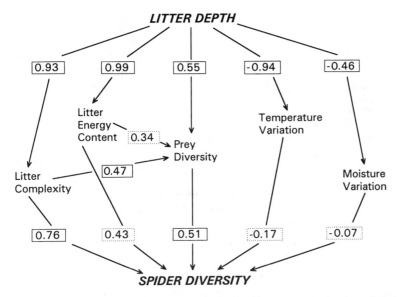

Fig. 7.9 Correlations between variables associated with litter depth, and their partial correlations with diversity of wandering spiders, for a field experiment in which the total amount of leaf litter was manipulated. Top row consists of simple correlation coefficients; other values are partial and multiple-partial correlation coefficients. Statistically non-significant values ($p > 0.05$) are boxed with a dotted line; all other values are significant at the 0.05 level or better. (After Uetz 1979.)

diversity the most. Litter complexity and prey diversity, however, were themselves significantly correlated. No data are presented on the groups of prey collected or how prey diversity was defined, so it is not possible to evaluate this connection any further. Uetz (1979) argues that by using seasonal totals to calculate the correlation coefficients, he made it more likely that litter complexity would show the strongest connection to spider diversity, because any effects of the other variables would be lost due to their greater variation within the growing season. When he analyzed the data separately for early, mid and late season, Uetz discovered that the strongest partial correlations with spider diversity were with prey abundance (0.32, Kendall's tau), litter complexity (0.54), and temperature range (0.78), for the respective times of the season. Although these relationships could not be assigned a level of statistical significance, the relative value of the non-parametric statistic indicates which factor is likely to be most influential at different times.

Uetz's first field experiment highlights the difficulty of untangling causal threads when a structurally complex component of the environment is manipulated. Correlation analyses are helpful in generating hypotheses about underlying mechanisms when many factors are manipulated simultaneously; however, such statistical searches do not constitute actual tests of these hypotheses, because the experimenter has not independently manipulated the variables in question.

Teasing apart the causes: two smaller-scale studies

Bultman & Uetz (1982, 1984) attempted to tease apart these causal threads by separately modifying two components of the litter: its nutritive content and its structural complexity. In his earlier studies Uetz utilized the relative amount of open space in the litter layer as a measure of structural complexity, a reasonable approximation since less-compressed litter ought to be more architecturally diverse. However, a relative measure of total amount of open space is not sensitive to heterogeneity in sizes and shapes of openings within the litter. Bultman & Uetz (1982, 1984) incorporated different degrees of this latter type of structural heterogeneity into the design of their manipulative experiment. In early April they determined the proportions of flat, bent, twisted or curled leaves in the litter of a beech–maple forest, and then created three experimental combinations of leaves that were either structurally less complex, equally complex, or more complex than the natural litter. Values for the measure of within-litter habitat space proposed by Uetz (1974) were 0.61, 0.75 and 0.86 for flat, natural and complex litter treatments, respectively. The investigators created similar treatments using artificial leaves of vinyl plastic, which provided no nutritive base for litter microarthropods. Collembola, which are major prey of litter-dwelling spiders, colonized the non-nutritive litter, but generally were higher in the natural-litter treatment (Bultman & Uetz 1984). Each treatment was replicated 24 times. The replicates were placed in 0.1 m² wire baskets arranged on the forest floor in a randomized block design. Litter depth in the baskets equalled that of the surrounding natural litter (4.4 cm). Each month for four months the litter from the baskets was removed, sieved for spiders and other arthropods that had colonized the litter, and was then returned to the baskets.

Structural complexity of the litter did not consistently affect the density of wandering spiders. In only one month (July) did a two-way ANOVA (complexity × nutritive value) yield a significant F value (3.17, $p < 0.05$) for the complexity treatment, and it was the natural, not the high, complexity treatment that had the highest abundance of spiders. Litter complexity did not influence species richness of wandering spiders in any month. These results differ from Uetz's interpretation of his earlier field manipulation (Uetz 1979), and strongly suggest that the major cause of increased species diversity in the first experiment was increased depth, not complexity, of the litter (Bultman & Uetz 1982).

Litter complexity had more of an effect on the density and diversity of web-spinning spiders, but statistically significant differences did not occur each sampling period. Thus Bultman & Uetz (1982) conclude that 'litter spatial heterogeneity appears a more important determinant of species richness in web-building than in hunting spiders but its effect is still minimal.' In three of the four months web-spinning spiders were more numerous in the artificial litter, a difference Bultman & Uetz (1982) attribute to the more rigid sites for attaching webs in the vinyl-plastic litter. Wandering spiders were more abundant in the natural litter in June and August, possibly because Collembola and other prey were more abundant, or because web-building spiders were less numerous (August only).

Bultman & Uetz's manipulative study makes it possible to evaluate mechanisms hypothesized to explain the positive correlations between litter depth and spider diversity uncovered in earlier studies. Apparently habitat volume is more important than the correlated variables of litter complexity and energy content in influencing the diversity of spiders, wandering spiders in particular. Bultman & Uetz point out, though, that the volume of the experimental treatments may be too small to detect an effect of litter complexity on densities of wandering spiders, because their high mobility makes it unlikely that they will remain within a basket for significant amounts of time. The same argument could explain the absence of strong effects of litter complexity on the species richness of wandering spiders. On a spatial scale of 0.1 m^2, litter quality had the larger effect, suggesting an aggregative numerical response by the spiders to higher densities of prey in the natural litter. Because population densities are higher in the deeper litter, Bultman & Uetz speculate that the larger habitat volume may allow vertical stratification and decrease rates of local extinctions.

Stevenson & Dindal (1982) performed a similar field manipulation of interstitial space and nutritive content of the litter, but with less clear results, primarily because of limitations in the experimental design. Rather than using mixtures of natural leaf shapes, Stevenson & Dindal employed flat and rolled freshly fallen maple leaves, formed by wetting and then pressing them to the desired shape. For non-nutritive litter they used filter paper disks. These two leaf shapes represent extremes of natural litter, making it difficult to relate the results of the experiment to the more subtle variations of natural litter. This is not a major flaw, however. The primary weakness of their experiment is the use of litter baskets so small (10 × 10 × 2 cm) that an average of fewer than three spiders colonized each container. In addition to being small, the boxes were constructed of metal hardware cloth with relatively small-sized mesh (0.64 cm), which would have prevented colonization by the larger litter-dwelling species.

Another aspect of their study that limits its generality relates to the composition of the spider community. The forest in which they performed the research has an unusually high population of the theridiid *Enoplognatha ovata* (pers. obs., Wise & Reillo 1985). Juveniles of this species overwinter in the litter, but eventually depart and spin webs under leaves of small saplings or brambles, remaining there through maturity and egg laying. Because Stevenson & Dindal's study area had an unusually high density of *E. ovata* adults, the proportion of litter spiders represented by immature *E. ovata* was not typical of most deciduous forest-floor communities. In fact, if one removes this transient species from the data and calculates colonization rates for species that live in the litter throughout their entire life cycle, it turns out that an average of only 0.8 spider colonized each litter box. This rate is an order of magnitude lower than that observed by Bultman & Uetz (1982), whose study was performed with litter baskets that were over 20× more voluminous.

Stevenson & Dindal conclude from a three-way ANOVA (leaf shape × leaf nutritive value × sampling date; including time as a treatment does not constitute pseudoreplication because the baskets were destructively sampled) that more species colonized the baskets with curled leaves; no indication is given whether or not nutritive condition of the litter affected species richness. Despite the correctness of the statistical analysis and the overall design, limitations imposed by the design of the experimental units and the type of litter employed limit the ability to extrapolate the results of this field experiment to natural leaf-litter communities.

Vegetation structure

Several experimenters have investigated whether or not availability of web sites limits populations of web-spinning spiders. The preceding discussion addressed this question for the litter layer, where the shape, arrangement and number of fallen leaves defines web sites for many species of small spiders. Web spinners that live in higher strata often seek sites on living and standing dead vegetation to attach their snares. Other features of the environment in addition to vegetation can afford attachment points for silk, but in most habitats some architectural feature of the vegetation is important. The following discussion will treat the broad question of web-site limitation of spiders that live above the litter layer.

Moor, heath and bog

Cherrett (1964) examined the distribution and abundance of spiders in an area covered primarily by peat supporting a mixture of moor vegetation. He attempted to associate spider dispersion patterns with major features of the environment. For example, Cherrett discovered that linyphiid abundance was correlated with the availability of arthropods across eight different vegetation types (Fig. 2.2), a relationship he conceded might reflect shared, direct effects of vegetation structure on spider and insect numbers rather than a numerical response by spiders to changing prey densities.

Cherrett also observed a distinctive pattern in the distribution of adults of two large and conspicuous orb weavers, *Meta merianae* and *Araneus cornutus*. These species, the only common orb weavers in the study site, did not occur on open moor, but were restricted to breaks in the 'blanket bog cover, i.e. erosion channels, stream edges, rock outcrops, and old mine workings.' Cherrett postulated that their distributions might be explained by variation in prey abundance, limited availability of sites for attaching webs, exposure of the webs to damage from the wind, or differences in moisture conditions. He discovered that potential prey were more abundant on the open moor than at typical web sites. Cherrett found no correlation between wind speed and the frequency with which the spiders rebuilt their webs; *c.* 80% of both species replaced their webs nightly. Preliminary preference experiments in the laboratory, and a reciprocal transplant experiment of very limited scope in the field, suggested that humidity might influence where the two species placed their webs within the preferred areas; however, Cherrett felt that the usually high humidity of the moors would not often be a limiting factor for either species. By eliminating all hypothesized factors but one, Cherrett concluded that both orb weavers concentrate in the channels and erosion areas because only these sites provide the needed attachment points for the supporting framework of the web.

Cherrett did not test his hypothesis experimentally by constructing artificial web supports in the open moor habitat. Colebourn (1974), in a set of experiments that in part owe their origin to Cherrett's earlier study, attempted such a test for another large orb weaver, *Araneus diadematus*. Colebourn studied this species on a limestone plateau covered by a thin cover of grass and patches of peat dominated by *Vaccinium* and *Calluna*. Webs of *A. diadematus* occurred most frequently in 'grikes', deep intersecting fissures in the bare limestone. In Colebourn's study area of 2550 m², 1100 m² (43%) was *Calluna* habitat, and the remainder was limestone pavement interlaced with grikes. Colebourn censused the area six times from June through September, locating a total of 4283 spiders (many spiders were of course counted several times; the maximum number at any single census was 1699, primarily instars II and III in late June). Of the seasonal total, 97% of the spiders had spun their webs in grikes. In one set of experiments, Colebourn modified the structure of the *Calluna* to determine whether or not the architecture of the heather was inappropriate for webs of *A. diadematus*. In one area of *Calluna* he cut into the vegetation 25 spaces, each 50 cm diameter by 10 cm deep, and then introduced 150 marked spiders (50 instar III, 50 instar V and 50 adults). As a control he introduced 150 marked spiders into 'similar *Calluna* areas where no spaces had been cut'. It is unclear whether or not the experimental and control sites were interspersed; if not, the statistical analysis is at least partially pseudoreplicated. Colebourn censused the area five days after the introductions. Of the 300 spiders introduced, 70 were found in webs on the heather; of these, 63% had spun webs in the cut-out spaces. The higher proportion in the manipulated vegetation resulted primarily from a strong preference for the openings by adults, but the numbers were small (14/50 and 2/50 of the introduced adults were found in the manipulated and control vegetation, respectively). Ninety-two of the 300 introduced spiders were found in nearby grikes, and the remainder (138) were lost. In a smaller-scale experiment, Colebourn introduced spiders on wooden frames into *Calluna* habitat (25) or grikes (15); after three days, more than half of the spiders had disappeared, with no statistically significant difference in rate of disappearance between grike and heath (73 and 56%, respectively). In a third study Colebourn placed spiders in separate 'gauze cells' in grike and heath situations, and found no statistically significant difference in survival after 20 days. This test was designed to uncover strong differences, as overall mortality was 63% (50% mortality in *Calluna*, 75% in grikes). Colebourn did not measure prey densities, but instead cites Cherrett (1964) as evidence that a difference in prey availability does not explain why *A. diadematus* is rare away from the grikes.

Colebourn's experiments suggest that the vegetation of the heath is not suitable for *A. diadematus*, but exactly which factors are important is unknown. Heather may not offer rigid enough supports. Survival of spiders on wood frames was low in the *Calluna* habitat, but was higher than the survival rate of spiders that had been introduced directly onto the vegetation. Colebourn suggests that adults spin their webs in grikes because openings in the *Calluna* are too small. Populations would become concentrated in the grikes if young spiders tend not to disperse far. Impetus for dispersal might be weak if web sites within the grikes were superabundant. Colebourn demonstrated experimentally that web sites are not limited in supply in the grike habitat by introducing spiders at two different densities. After three weeks survival was density independent, and final densities were higher than in non-disturbed areas of the grikes. Thus the absence of vegetation adequate to support webs of adults in otherwise habitable areas explains the dispersion pattern of *A. diadematus* on one scale. On a smaller scale, within suitable habitat patches, factors other than a shortage of adequate web sites limit densities of this orb weaver.

Floronia bucculenta is a linyphiid that spins its sheet web in low, dense vegetation. Schaefer (1975) counted an average of three occupied webs/m^2 in grassy vegetation on a dry *Sphagnum* bog. He tested the hypothesis that suitable web sites were limiting this population by introducing artificial web sites of wire-netting formed into 10 cm diameter cylinders. In this habitat the area of the sheet portion of the web in natural vegetation approximates the cross-sectional area of the cylinder. Schaefer located 11 open 1 × 1 m plots and increased the density of web sites in six of the plots by placing 20 wire cylinders in each plot; the remaining five plots served as controls. To each plot he introduced 10 female *F. bucculenta* and after 2.5 weeks determined the density of females in webs. Mean density on the control plots (± s.e.) was 3.2 ± 0.8, but on the experimental plots had increased to 10.2 ± 0.8. Approximately 70% of the introduced spiders had remained and built webs in the plots with additional web sites, but most of those added to control areas apparently had vanished. Because spiders were not marked it was not possible to detect web invasions, which may have been frequent because available web sites were clearly in short supply. The experiment provides clear evidence of web-site limitation in this population.

The sagebrush community: direct and abstract approaches
Experiments are most straightforward whenever habitat units to be manipulated are discrete and numerous, such as occurs in high-desert habitats dominated by the big sage *Artemisia tridentata*. Hatley & MacMahon (1980) manipulated the structure of big sage bushes in order to determine the influence of plant architecture upon the structure of the entire spider community, which they divided into five guilds: nocturnal hunters (clubionids and similar spiders that hide in daytime retreats), runners (philodromid crab spiders), ambushers (sit-and-wait crab spiders, the thomisids), agile hunters (salticids and oxyopids, spiders that rely on keen eyesight to locate and pursue prey) and web builders. Their study site was a one-hectare plot divided into 25 sub-plots of 20×20 m. Sub-plots were assigned randomly to three groups and on each sub-plot 50 shrubs were selected at random to receive the treatment assigned to that sub-plot. On eight sub-plots 50% of the foliage on each bush was clipped and on another eight sub-plots branches were tied together in order to increase foliage density. Bushes on the remaining sub-plots served as controls. Hatley & MacMahon altered the shrubs in April and then removed spiders from five shrubs on each sub-plot weekly June through August, and once in the beginning of October. Seasonal means were used in analyzing the results.

Hatley & MacMahon report that 'tied shrubs had significantly more resident species and guilds than either control or clipped shrubs. No significant differences were found between clipped and control shrubs.' The Wilcoxon ranked sum statistic was used to test for differences, but no values are given. Surprisingly, values of species numbers for the three categories (clipped, tied and control) are not presented, so the magnitude of the response is unknown. Also surprising is the authors' decision to abandon the original experimental design for most further data analyses. For example, they searched for correlations between aspects of spider community structure and characteristics of individual bushes. This was possible because they had determined the foliage density of sampled shrubs by photographing each against a gridded background of 20 cm squares. Hatley & MacMahon calculated the correlations between three measures of diversity [S, H' and J' (H'/H'_{max})] and several parameters (shrub height, shrub volume, foliage mass at different heights, percent dense foliage, shrub foliage diversity and others). Of 30 correlation coefficients calculated for all the species sampled on a bush (i.e. including juveniles of species which may not have remained on the bush until adulthood), 16 were statistically significant. When spiders in early juvenile stages were eliminated, 24 correlation coefficients were significant. Strongest correlations were with height or volume of the bush, but the correlation never explained more than 10% of the variance in S, H' or J'. Correlation coefficients for shrub foliage diversity ranged from 0.17 to 0.27 (only 3 to 7% of the variance). Data for individual shrubs from the same sub-plots were treated as individual data points; however, shrubs within a sub-plot received the same experimental treatment, making many of their foliage measurements similar. Thus data on shrubs from the same sub-plot are not independent, which violates an assumption of statistical analysis. Furthermore, some shrubs were somewhat arbitrarily defined to be discrete units, as they were as close as 10 cm to each other. Though problems of pseudoreplication exist, the inflated degrees of freedom in the correlation analyses do not obscure the fact that relationships between indices of spider community structure and vegetation characteristics were weak.

Hatley & MacMahon's decision to look for patterns by calculating correlation coefficients is not unreasonable, but what is most surprising about the rest of their analysis is the decision to re-assign bushes to three new categories in order to perform ANOVA. New groupings are based upon percentages by volume of canopy, open foliage and dense foliage. Clipped, tied and control bushes all appear in the three new categories, but in different proportions. One wonders if the investigators decided to re-group the treated bushes because ANOVA's based on the original design did not yield many statistically significant results. The new ANOVA yields significant F-values, with the mean number of spider species being higher in the two categories that have greater numbers of tied shrubs. However, because the original experimental design has been abandoned it is difficult to interpret the results. For example, the new categories II and III have nearly identical numbers of species (mean of 5.1 per shrub for both categories), yet 55% of the shrubs in category III had tied branches, whereas only 31% of the shrubs in category II were treated in this manner.

Hatley & MacMahon's study is a properly designed manipulative experiment. Given the strengths of such an approach, it would have been best to calculate a mean value of S, H' and J' for each sub-plot and then analyze the pattern using a one-way ANOVA that adhered to the original experimental design. Calculating the mean value of the shrub structural parameters for each sub-plot and then comparing these would have been an appropriate way to uncover likely mechanistic explanations for significant effects uncovered by the original ANOVA. This is an excellently designed study whose results unfortunately were not analyzed in the most straightforward manner. Nevertheless, the outcome appears clear: experimental manipulation of the vegetation affected spider community structure, but perhaps not to the extent that one would have predicted.

Robinson (1981) conducted an experimental study of the spiders on Hatley & MacMahon's study area using a more abstract approach. Instead of modifying natural vegetation, he introduced different types of artificial substrates and documented their colonization by spiders resident on the natural vegetation. Unlike Schaefer's (1975) experiment with artificial web sites, Robinson's goal was to identify aspects of plant architecture to which spiders respond, rather than to discover whether or not the availability of suitable plant structure limits spider densities. Colonization of Robinson's experimental units is not proof that foraging sites or retreats are limited in the habitat, because densities in the entire area were not measured. By examining all categories of spiders on his artificial vegetation, Robinson, like Hatley & MacMahon (1980), did not limit his focus to the role of plant structure as sites for web attachment.

Robinson's experimental units were cubes of 2.5-cm-mesh chicken wire, 30 cm on a side, that supported one of six arrangements of strands of macramé jute. Replicates of the different patterns of jute scaffolding were left in the field from 1 to 16 days. Spiders that had colonized the units were collected by banging the unit against the sides of a large extracting funnel. The type of jute scaffolding influenced the number of individual spiders and the number of species appearing on the units after 16 days ($p(F) < 0.001$, one-way ANOVA; d.f. = 5, 102). The design that had the most jute (a dense arrangement of strands in three dimensions) had the highest mean number of individual spiders (5.7) and spider species (3.4). Cubes with more widely separated strands that were arranged vertically had the smallest number: 2.5 individuals and 1.6 species per structure. Using non-parametric tests to compare the utilization of different types of cubes by the more abundant species, Robinson found some preferences for either vertical or horizontal strands, or strands that were more or less widely spaced. The design and statistical analysis are clear and correct. The experiment demonstrates that some spiders respond differentially to a variety of artificial structures placed in the field. Unfortunately, the connection between the behavior of spiders on jute and their responses to the architecture of natural vegetation is obscure, and was not addressed by the investigator. Thus the relevance of this field experiment to the dynamics of spider populations on natural vegetation is questionable.

Substrate or prey availability?

Densities of web-spinning spiders may change in response to variation in prey densities; earlier chapters evaluated evidence supporting this generalization. Correlative and experimental studies discussed in this chapter have demonstrated that architecture of the surroundings, both inanimate and living, also affects the density of those species that must find supports from which to suspend their snares. Rypstra (1983) performed an experiment in field cages to test simultaneously the influence of these two components of the environment. After releasing 100 spiders into large ($3 \times 3 \times 2$ m) completely screened enclosures located in the understorey of an oak forest, she counted the number of spiders that built webs as a function of the availability of artificial substrate (number of wooden orange crates; 0–4) and rate of supply of prey (*Drosophila melanogaster*; rates varied from 200 to 3000 flies/day). Both variables strongly influenced the number of spiders building webs and coexisting in the cages a week after having been introduced (Fig. 7.10). Spiders that vanished had been eaten by other spiders.

The sheet-web spinners showed a less rapid response to increases in prey than did the orb or tangle-web weavers. With one orange crate and 3000 flies a day, the number of sheet-web spiders levelled off at about 30 per cage, the number in the cage with four orange crates and 200 flies per day. In contrast, the number of spiders of the other two guilds, with one orange crate and the highest prey numbers, was twice the number of spiders coexisting on four orange crates and the lower rate of prey supply (Fig. 7.10). The tendency of sheet-web spiders to be less responsive to food supply may reflect their greater investment of energy in the web, and/or more-specific requirements for web supports. Differences between orb and sheet spinners observed by Rypstra are consistent with the generalizations and preliminary empirical findings of Janetos (1982a,b; Chap. 2).

Rypstra's experiment was not a complete factorial design, because feeding rate was held at 200 flies per day for manipulations of substrate availability, and prey feeding rates were varied in enclosures with a

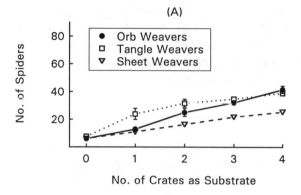

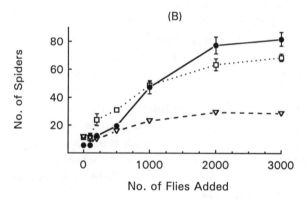

Fig. 7.10 Influence of substrate availability and prey abundance upon the number of spiders in 3 × 3 × 2 m screened enclosures six to eight days after 100 spiders had been introduced. A preliminary experiment with a common theridiid (the house spider *Achaearanea tepidariorum*) revealed that the number of spiders fell to the carrying capacity of the cage (with one orange crate and 200 flies (*Drosophila*) provided per day) within a week. Individuals from several species of each guild (orb, sheet or tangle-web weaver) were added together, but all spiders were within 10% of each other in length. Means ± s.e. based upon two replicates – one from 1979, the other from an experiment run the following year. Many standard errors are less than the symbol width. (A) Response of spiders to number of orange crates placed in each enclosure. Each cage received 200 fruit flies per day. (B) Effect of prey availability upon number of coexisting spiders. Each enclosure contained one orange crate. (After Rypstra 1983.)

single orange crate. A completely crossed design would have revealed information on the interaction between substrate availability and prey abundance. This objective could have been achieved within the logistical constraints of the study by decreasing the number of different levels of each treatment while retaining the range of each.

Rypstra's experiment reveals the responsiveness of web spinners to changes in two major components of their environment, but the results do not bear directly on the actual influence of substrate or prey supply upon these species in the world outside the screened enclosure. This is not a criticism of the study, because her goal was not to determine whether a particular population was food or substrate limited. As Rypstra (1983) points out, her most intriguing finding is the observation that when prey were sufficiently dense, three species of tangle-web weavers and two orb weavers changed their mode of web construction and their foraging behavior: 'Individuals moved freely through each others' webs and fed on the prey caught by other individuals.' Tolerance of other spiders typifies colonial and social spiders. The discovery that non-colonial species exhibit this tolerance at unusually high prey densities supports one of the popular hypotheses for the evolution of colonial and cooperative social behavior in spiders, i.e. that it evolved as a consequence of spiders aggregating at dense pockets of prey.

Silk of colonial and social spiders provides attachment sites for the webs of other spiders in the colony. In fact, one advantage of coloniality in spiders is the ability to spin webs across spaces unavailable to a single individual. In general, cooperative web building may offer advantages whenever the gain per individual in terms of increased prey capture is greater than the negative costs of living close to competitors. Other gains from grouping are also feasible. Several investigators have reviewed patterns of colonial and social behavior among spiders and the hypotheses proposed to explain its evolution (e.g. Shear 1970, Kullmann 1972, Buskirk 1981, D'Andrea 1987, Uetz 1986a, 1988). This topic is complex and fascinating. It merits further examination by any reader intrigued by how spiders provide anchoring points for each other's webbing, and how some spiders cooperate in prey capture and rearing of young.

Abiotic anchors

Vegetation, both living and deceased, is a major but not sole component of the spider's structural environment. For example, the soil (itself formed partly of decayed plants) is substrate for numerous burrowing spiders. The abandoned burrows of mammals support several species of web spinners (Heidger 1988). Other examples of non-vegetative architectural components of the spider's world have been discussed earlier: e.g. rocky outcroppings as sites of web attachments; depressions in the soil of desert grassland; the aquatic environment for spiders that hunt on its surface or live submerged beneath; the protective layer of snow for winter-active spiders. Clearly, many abiotic aspects of the environment are important for spiders, as shown by both correlative and experimental studies (Chap. 5).

And of course one must not overlook the entire sum of human architectural achievement, the abode of uncounted hordes of well-adjusted house spiders and their more transient relatives. This is one niche ecologists have left virtually unexplored.

A call for multi-faceted approaches

Most of the investigations reviewed in this chapter have focused on whether or not a particular feature of the non-trophic web influences spider distribution or density. In their singularity of purpose these studies resemble most investigations of biotic interactions, in that food supply, or competition, or the role of natural enemies, is investigated as a factor in isolation from others. To do otherwise usually requires extensive, long-term research. Rypstra's (1983) cage studies on the contribution of substrate availability and prey levels to levels of cannibalism is an example of a study that begins to integrate explicitly the abiotic and biotic factors. However, it is basically a laboratory study. Riechert's research, spanning almost two decades, has treated the roles of vege-tation, physical factors, food supply, environmental unpredictability and territoriality in the dynamics of different populations of one species. A community study in the natural environment that combined the scopes of these two examples would be a formidable challenge. Under-standably, most community-level research with spiders has focused on one or a few driving variables that potentially influence community structure, which is often defined in terms of a simple index. More emphasis in future research should be placed upon understanding how two or more processes interact to affect the dynamics of spider popula-tions within the complete complexity of their ecological webs.

Synopsis

Vegetation structure and properties of the leaf litter anchor the maze of trophic connections that forms the metaphor of the ecological web. Early investigators recognized that the physical structure of the habitat (its physiognomy) could profoundly affect the composition of the spider community. Effects arise not only from variations in the availability of supports for anchoring the snares of web-building species, but also from the provision of retreats and modification of the microclimate, which impact spiders in numerous families. Initial evidence came from correlations between the types of spiders inhabiting an area and both the structure of the living vegetation and the composition of the plant debris that forms the litter layer.

Early studies of successional communities uncovered regularities in the association of particular combinations of spiders with the overall plant community. More-recent quantitative approaches have uncovered correlations between (1) structural diversity of the vegetation and the species diversity of web-spinning spiders; (2) amount of vegetation and the activity/abundance of web spiders; (3) density of foliage on conifer branches and the size-distribution of spiders; (4) presence of flowering herbs and densities of spiders foraging for pollinating insects; (5) nature of the substrate and the microhabitat distribution of particular species of wandering spiders; (6) depth and complexity of the leaf litter and the species diversity of the cursorial spider community. In several instances mechanisms underlying the influence of the vegetation seem apparent: i.e. provision of shade in stressful thermal environments; furnishing hiding spaces for retreats; providing protection from the wind; attracting prey; furnishing attachment sites for the snares of web weavers.

Correlations between variables in non-manipulated systems suggest causal connections, but require independent confirmation, which comes most directly from controlled field experiments. In addition, experimentation provides the opportunity to identify which aspects of the environment's physiognomy directly affect spider populations. For example, the amount of leaf litter could influence spider communities through several routes: influencing the productivity of prey populations, modifying the microclimate, providing spaces for webs and retreats. Separating these components through experimentation is not straightforward. Some experimenters have attempted to distinguish contributions of nutritive content, structural complexity and total volume of leaf litter. Results suggest that, in terms of effects upon spider density and diversity, litter volume *per se* is more important than structural complexity or energy content of the litter. More research is clearly needed. The experiments conducted so far have been carefully designed, but it is difficult to draw strong conclusions from some of them because of problems of scale or replication.

Investigators have introduced artificial models of vegetation or have directly manipulated the structure of living plants. Early experiments provided evidence that a shortage of suitable sites affects densities of particular species. Later research manipulated the structure of shrubs, or introduced artificial shrubs of different design. Both approaches have yielded significant effects of plant architecture upon spider communities, but it is difficult to extrapolate from the results of these studies to the type and magnitude of influence that natural variation in vegetation structure exerts upon spiders.

Correlative and experimental studies have uncovered clear influences of vegetation structure, composition of the leaf litter, and other aspects of the environment's physiognomy upon spider populations. Most experimental studies have manipulated components of the structural environment in relative isolation from the impact of biotic interactions within the spider's ecological web. A notable exception, although a laboratory-type study because entirely enclosed field cages were used, is an experiment that explicitly examined the joint effects of web-site availability and prey abundance upon rates of cannibalism.

8 · *Untangling a tangled web*

Introduction

Readers not yet weary of the metaphor may realize that the first six chapters have examined gentle pluckings of the spider's ecological web: field experiments designed to probe the strength of threads connected directly to neighboring actors in the ecological drama. Experimental evidence has been evaluated for food limitation and intraspecific competition, competition between spider species, limitation of spiders by natural enemies, and the impact of spiders upon their prey populations. Most of these interactions yield what conventionally are termed *direct effects*, though exploitative competition is actually an indirect interaction mediated through impacts on shared prey populations. The last chapter expanded the scope by examining how other components of the environment influence the biotic interactions that have formed the threads of the metaphor. Now it is time to probe consequences of persistent plucking and poking of more-distant regions of the web. What do field experiments reveal about the larger set of connections, links between spiders and other taxa of predators, and organisms on trophic levels several energy-transfer steps away? In short, what do field experiments reveal about *indirect effects* in complex communities of which spiders are an integral part?

Indirect effects in a simple system

What community could be simpler than two spider species and their prey? David Spiller uncovered both exploitative competition for prey and interference competition in the ecological web comprised of the orb weavers *Metepeira grinnelli* and *Cyclosa turbinata* and their prey in a salt marsh (Spiller 1984a,b; Chaps. 4,5). This system may appear simple, but Spiller (1986b) discovered that predicting how prey populations respond to changes in densities of the two species can become complicated.

During June the estimated total consumption rate of small insect prey by orb-weaving spiders (number of 1–2 mm prey consumed/m³/day) was significantly higher on the *Metepeira*-removal plots, where only *Cyclosa* remained, than on the control plots, where both species were present (2.53 ± 0.36 versus 0.96 ± 0.37, respectively; mean $\pm$ s.d.; $p < 0.05$, Tukey method for pairwise comparisons). More small prey were captured in plots without *Metepeira* because *Cyclosa* densities were twice as high as a result of reduced interference from web invasions by the larger *Metepeira*. Although in June *Metepeira* consumes some small insects, it feeds primarily on larger prey, so that its presence or absence does not directly affect the mortality rate of small insects; however, the indirect effect of *Metepeira*, expressed through *Cyclosa*, is clear. Because Spiller (1984b) uncovered intraspecific exploitative competition for prey among *Cyclosa*, it is reasonable to predict that the higher density of *Cyclosa* should lead to a lower overall abundance of smaller insects. Thus a reduction in *Metepeira* densities may actually have caused a decrease, not an increase, in the abundance of small insects. Spiller suggests that this hypothesis be tested by conducting a competitor-removal experiment with larger plots than he used.

Indirect interactions involving spiders may be widespread. If ubiquitous, such indirect effects will most likely be found to result less from competition than from direct predation; competition between spider species appears to be much less widespread than predation by spiders on other spiders (cf. Chaps. 4,5). Polis *et al.* (1989) termed predation between species of the same predatory guild 'intraguild predation (IGP),' a phenomenon they argue is more prevalent than many ecologists realize. Intraguild predation is undoubtedly a major component of the spider's ecological web.

A predatory persona of many roles

Because spiders are generalists that consume many of the same prey species, including each other, they potentially affect each other indirectly through exploitative competition or directly through predation. Cannibalism and interspecific predation probably occur more frequently among wandering spiders than web builders, because the former are more apt to encounter each other as they forage for prey. Negative interactions uncovered to date between lycosid species are most readily interpreted as IGP (Schaefer 1972, 1974, 1975). Several ecologists have suggested that cannibalism and IGP keep densities of wandering spiders low enough to prevent exploitative competition. Competition for prey may occur among wandering spiders [e.g. preliminary evidence of exploitative competition among young *Schizocosa ocreata* (Wise & Wagner 1992; Chap. 5)], but the experimental evidence is too scanty to justify generalizations about the relative magnitude of exploitative competition versus IGP in communities of wandering spiders.

Most species of web builders do not regularly prey upon other spiders. Web invasions by other web-spinning spiders usually result in agonistic interactions followed by retreat of one member of the pair (e.g. Riechert 1978b, 1979, Wise 1983, Hoffmaster 1986). Web builders, however, are not immune to predation from other spiders. Many species routinely invade webs and prey upon the owner [e.g. *Argyrodes* spp. (Theridiidae); several salticid species; the entire family of pirate spiders (Mimetidae), which are spider specialists; cf. Chap. 5].

Spiders in most communities belong to a large guild of generalist predators that includes other invertebrate taxa, which may make it difficult to predict how a change in spider abundance will affect prey populations. For example, in evaluating whether or not two species of lynx spiders (Oxyopidae) limit densities of cotton pests, Nyffeler *et al.* (1987a,b) emphasize that the diets of these spiders include predacious invertebrates, many of which probably eat the pest species. The net impact of spiders upon pest populations conceivably could be zero or even positive, depending upon the relative vulnerability of the pest and its natural enemies to spider predation.

The web of indirect effects in which spiders are entangled stretches farther than ramifications of intraguild predation. Spider predation upon herbivores and saprovores could have consequences for the dynamics of plant populations and rates of decomposition. And spiders need not always be viewed as owning the leading roles – perturbations in other regions of the web could affect spider populations through changes in resource levels or pressure from natural enemies.

In the following pages intraguild predation will be examined first, followed by a treatment of evidence for cascading effects in grazing and detritus-based food chains. In order not to become entangled in a maze of terminological confusion, and to avoid being trapped into unreservedly espousing the virtues of field experimentation, a brief digression comes first: What is meant by the *ecological community* and what are the rewards and risks of perturbing *complex communities* in field experiments?

Experimenting with complex communities
Spider communities and communities with spiders

It is a disturbing realization, usually repressed by ecologists, that a community is practically any species collection that a community ecologist considers worthy of study. This gnawing reminder of the softness of their discipline periodically surfaces when ecologists gather together in symposia to assess the state of their science. McIntosh (1987) reviews some recent efforts, commenting that 'a continuing problem of community ecology, reductionist and otherwise, is the difficulty of transcending familiar taxa and of defining community.' Introductory textbooks define a community to be a collection of species inhabiting an arbitrarily defined region, though often ecologists further restrict the collection along taxonomic or trophic lines, and may delimit the region to a habitat characterized by a type of vegetation or an easily recognizable physical attribute. Implicit in the notion of community is the importance of species interactions; if the degree of biotic connectedness is unknown, ecologists sometimes substitute more neutral terms, such as 'association' or 'assemblage.' Resource relationships are emphasized by using functional terms such as 'guild' or 'herbivore community.' Terms such as 'food chain' or 'food web' substitute for 'community' when broader, more-complex functional relationships are the focus. A recent textbook has expanded the options by eschewing traditional hierarchies and refusing to recognize a distinction between the community and ecosystem level of organization, preferring instead to emphasize that the studies of ecological energetics and nutrient recycling 'represent supplementary means of approaching an understanding of community structure and functioning' (Begon, Harper & Townsend 1990).

The underlying concern that fuels debates over definitions of the ecological community is the shared desire of ecologists to generalize about patterns of interactions between species, i.e. to formulate principles of community structure and functioning. Thus discussions of what is a community are healthy as long as participants realize that the goal of their dialogue should be to find the best way to generalize about ecological phenomena, and not to characterize some elusive mythological construct. The actual meaning of 'community' depends upon the context in which it is used, and it is in this spirit that the term has been employed throughout the first seven chapters. Most frequent in earlier chapters is the term 'spider community,' a usage that is most compatible with the outlook that interspecific competition is a major interaction determining relative abundances of species. Recent chapters have expanded the definition to emphasize direct interactions between spiders and trophic levels below and above; thus spiders have been viewed as components of larger communities, such as the 'forest-floor community.'

Dismissing the problem of defining 'community' is relatively easy; what is more challenging is interpreting the results of field experiments with broadly defined communities, complex systems that include almost uncountable direct species interactions and inevitable indirect effects. Experimentally perturbing complex communities is a relatively new approach to the study of community organization and function. It is an approach that offers promise of uncovering indirect interactions in the ecological web.

Field experiments and indirect effects in complex webs

Field experiments often are the most convincing vehicle for distinguishing between alternative hypotheses. Nevertheless, acceptance of field experimentation as a primary approach in community ecology is by no means universal (e.g. Cody 1974, Gilpin, Carpenter & Pomerantz 1986). Early warnings against over-emphasizing field experimentation were based on the realization that simple removal experiments can potentially yield misleading results because of unrecognized indirect interactions (e.g. Levins 1975, Levine 1976, Holt 1977, Davidson 1980, Levins & Lewontin 1980). However, this criticism applies only to field experiments analyzed in isolation; they must always be interpreted in light of other information about the manipulated system (e.g. Reynoldson & Bellamy 1971, Bradley 1983, Bender et al. 1984, Wise 1984a, Abrams 1987a). Furthermore, field experiments offer a powerful tool for actually detecting indirect interactions in complex communities.

Long-term experiments in which the experimental perturbation is maintained can reveal effects of indirect interactions between species in the ecological web. Bender et al. (1984) term this type of perturbation a 'press experiment,' which provides information on both direct and indirect effects. It is the type of field experiment most frequently employed by field experimentalists. A single, short-term perturbation – a 'pulse' experiment – is usually limited to revealing direct effects only. Although better suited for detecting indirect effects, press experiments are not immune to problems (Bender et al. 1984, Yodzis 1988). For instance, numerous manipulations are required to estimate all pairwise interaction coefficients needed to unambiguously separate direct and indirect effects. Lumping species into broad functional categories is one partial solution to this logistical dilemma. Care must be taken in selecting categories, because including species that are functionally quite different, or failing to recognize all members of a functional group, may lead to erroneous conclusions about mechanisms underlying a system's response to a perturbation. This caveat, however, applies to all approaches to studying ecological communities, because broadly defined categories can only approximately reflect functional groupings within complex systems.

Another problem with using press experiments to reveal indirect interactions is the amount of patience demanded. Formal analysis of press experiments indicates that a new equilibrium must be reached before one can make final inferences about indirect effects (Bender *et al.* 1984, Yodzis 1988). Attainment of equilibrium may require a long time (e.g. Brown *et al.* 1986) – several cycles of a typical research grant. Inevitable logistical constraints thus dictate that results of perturbation experiments be interpreted cautiously. One aid to interpretation is 'supplementing ecological experiments with the fullest amount of descriptive natural history' (Bender *et al.* 1984). This admonishment parallels McIntosh's (1987) advice that field experimentation is most powerful when combined with other approaches, particularly when experimenting with complex networks.

Not all of the experiments to be discussed in this chapter fall neatly into the 'press/pulse' dichotomy, particularly cage experiments in which animals are added but densities are not maintained at the artificial level throughout the experiment. Some experiments are of such short duration that the treatment appears to be a press perturbation, but not enough time elapses to detect indirect effects resulting from the manipulation. Nevertheless, the distinction of Bender *et al.* (1984) is a useful way to categorize and evaluate manipulative studies with complex communities.

Intraguild predation (IGP)

Separating effects of competition for limited resources from IGP requires a combination of manipulative field experiments and detailed observations of feeding behavior. Such research has been conducted with two systems involving spiders and other taxa: spiders, scorpions and solpugids (three different orders of Arachnida) in a desert community (Polis & McCormick 1986); and spiders and lizards on Caribbean islands (Pacala & Roughgarden 1984, Spiller & Schoener 1988, 1990a). Hurd & Eisenberg (1990) performed a short-term pulse experiment with a third system, wolf spiders and mantids, that also bears directly upon IGP. Aspects of some of these studies are discussed in the treatment of spiders' natural enemies in Chapter 5.

Desert spiders and their arachnid antagonists

Scorpions, spiders and solpugids are conspicuous predators in desert communities. Polis & McCormick (1986) studied interactions between these arachnids by measuring dietary overlap, including the extent of IGP; and by performing a long-term removal experiment in which they reduced densities of the most abundant scorpion, *Paruroctonus mesaensis*.

Diets overlap substantially between the three arachnid groups. For example, Polis & McCormick report that *P. mesaensis* preys upon 23 of the 32 species taken by the spider *Diguetia mohavea* [Diguetidae; a small family whose members construct tangled webs less than a meter off the ground under vegetation in desert habitats (Kaston 1978)]. Schoener's (1968) index of dietary overlap was 49%. All 18 species of prey observed in the webs of the black widow spider, *Latrodectus hesperus,* are also prey of the scorpion *P. mesaensis*. Dietary overlap between several solpugid species and *L. hesperus* and *P. mesaensis* is also extensive.

IGP between these arachnids is also high. Polis & McCormick cite other studies demonstrating that solpugids and spiders prey upon each other as well as the other groups, but they argue that most of the IGP in their system involves scorpions preying upon spiders, solpugids and other scorpions. Most of this predation involves *P. mesaensis*, which comprised ≥ 95% of the scorpions in their study area. During four years of observations, in which they documented over 900 instances of predation by *P. mesaensis*, Polis & McCormick found that 16 of the 39 spider species on their plots comprised 8% of the scorpion's diet, and all 11 solpugid species in the community accounted for 14% of its diet.

Polis & McCormick conducted a 29-month press manipulation in order to determine whether *P. mesaensis* influences the densities of spiders and solpugids in the community and, if so, whether responses result from a reduction in IGP, exploitative competition, or both. The design employed 300 removal plots and 60 control quadrats, all 10×10 m, 'randomly established in nearby, similar, but unmanipulated areas.' Starting in June 1981 and continuing through October 1983, the

Table 8.1. *Abundance and size of spiders in the* Paruroctonus mesaenis-*removal area and the adjacent control area at the end of a 29-month field experiment. The most abundant scorpion had been removed to test for effects of IGP and exploitative competition with spiders. Means ± s.d.*

	Date	Removal area	Control area	t	d.f.	p
No. webs/m²	Spring 1983	2.18 ± 2.27	1.62 ± 1.69	2.80	398	0.005
No. spiders/	Spring 1983	2.37 ± 1.65	1.29 ± 1.67	2.48	58	0.01
pitfall trap	Autumn 1983	2.97 ± 1.59	2.93 ± 2.02	0.07	58	n.s.
	Spring 1984	6.40 ± 4.95	3.00 ± 2.03	4.98	28	0.001
Spider length	Spring 1983	3.14 ± 1.64	3.37 ± 3.38	0.29	108[a]	n.s.
	Autumn 1983	2.58 ± 1.96	2.74 ± 1.49	0.60	175[a]	n.s.

Note:
[a] Estimated from mean number/pitfall trap and number of traps in each area.
Source: Polis & McCormick 1986.

removal plots were visited on 61 nights, and over 6000 scorpions were removed. Polis & McCormick (1986) do not present data on initial densities of *P. mesaensis*, but in another publication (Polis & McCormick 1987) report that initial densities of two other scorpion species did not differ between the removal and control areas. The average density of all scorpions in the region of the study was 1879/ha (Polis and McCormick 1986), which yields a rough estimate of 5355 *P. mesaensis* on the removal area at the start of the experiment (1879/ha × 0.95 × 30 000 m² ÷ 10 000 m²/ha). Therefore the number removed during the experiment exceeds the number estimated to be present initially. It is not clear to what extent the population density of *P. mesaensis* actually was reduced in the removal area; however, densities of scorpions clearly differed between removal and control areas at the end of the experiment (October 1983). After a week of nightly surveying 60 quadrats in each area, Polis & McCormick concluded that there were 20 × more scorpions active per square meter on the control area, compared to the removal. Mean numbers per 60 quadrats, ± s.d., were 237 ± 70 and 13 ± 11, respectively.

At the end of the experiment the number of spider webs, and the number of spiders caught in pitfall traps, were higher on the removal area, compared with the control (Table 8.1). The number of solpugids trapped on the two areas did not differ. There was no evidence of competition for prey between *P. mesaensis* and spiders: mean spider length (Table 8.1), mean solpugid length, prey capture and reproductive parameters of the spider *Diguetia mohavea*, and the abundance of potential non-arachnid prey, did not differ between removal and control areas at the end of the experiment. This pattern of statistically significant and insignificant differences led Polis & McCormick to conclude that predation by the abundant scorpion *P. mesaensis* lowers spider densities, but has no effect on solpugid numbers.

The investigators mention the possibility that scorpions could have been depressing prey numbers, thereby inducing exploitative competition with spiders, but that the expected increase in prey numbers resulting from release from scorpion predation went undetected because spider numbers also increased in the scorpion-removal plot. Polis & McCormick argue that this explanation is unlikely, because the increase in spider biomass was <5% of the biomass of scorpions removed. They also raise the question of why solpugid numbers did not increase, given that spider numbers did. Solpugids form a larger fraction of *P. mesaensis*'s diets than do spiders. The influence of indirect interactions is one explanation: increased numbers of spiders in the removal plot may have suppressed any increase in solpugid numbers resulting from lowered rates of predation by scorpions. Polis & McCormick argue that 'a more parsimonious explanation focuses on behavioral differences.' Because solpugids wander much more than scorpions, 'movement and subsequent mixing of the population may make it difficult to detect any changes in the removal, as compared with the control, quadrats.'

The long duration of the manipulation and the use of extensive, open areas are two major strengths of Polis & McCormick's study. Unfortunately, however, the decision to use a large removal area led to a non-replicated design. The 300 removal quadrats were contiguous subdivisions of one 30 000 m² area (Polis, pers. comm.). Thus removal and control plots were not physically interspersed, which violates a major requirement of experimental design (Hurlbert 1984). Because the 10×10 m quadrats are sub-samples of a single experimental area, using statistics to compare densities of spiders between removal and control areas as a test of the influence of *P. mesaensis* constitutes pseudoreplication. Spider densities clearly differ between the two areas, but it is not possible to conclude that altered predation pressure from scorpions has produced the difference. In fact, densities of spiders on the two areas are remarkably similar, considering the $20 \times$ difference in scorpion density. A more parsimonious explanation is that scorpions have no effect, or at best a very minor influence, on spider numbers. However, even this circumspect conclusion is unjustified. Without replication one is in the unfortunate predicament of not being able to commit either Type I or Type II error. To conclude no effect of scorpion numbers in a non-replicated design would also be premature.

Orb weavers and lizards

Pacala & Roughgarden (1984) discovered both direct and indirect effects of lizards on the abundance of arthropods in a field experiment designed to determine whether two species of anoles depress arthropod numbers (details of their experimental design are discussed in Chap. 5). Arthropods on the forest floor in the center of the enclosures were more abundant in the plots with reduced densities of lizards; mean numbers per sticky trap ($\pm$ s.e.) were 26.3 ± 2.7 and 12.5 ± 0.5 for low- and high-density lizard plots, respectively [calculated from the mean number per trap for each replicate from Table 2 of Pacala & Roughgarden (1984); $p < 0.05$, Kruskal–Wallis test, d.f. = 4; non-parametric test used because variances are significantly different]. An increased number of dipterans produced the significant effect. The anoles manipulated by Pacala & Roughgarden forage primarily on the forest floor, and dipterans are commonly found in their guts. Thus the difference between lizard treatments in abundance of forest-floor arthropods results from different intensities of lizard predation, a direct effect.

In contrast to the pattern on the forest floor, arthropod densities in the understorey were higher, not lower, in the plots with more lizards (Fig. 8.1). These plots had more Coleoptera, Diptera and Homoptera, which are common prey of orb-weaving spiders. The three most conspicuous web-building spiders were over an order of magnitude less abundant in the plots with greater numbers of lizards (Chap. 5). It appears that release of spiders from lizard predation caused spider numbers to increase to levels that depressed the abundance of flying insects. This indirect effect of a reduction in lizard numbers on insect prey is analogous to that observed by Spiller (1986b) when he reduced numbers of the orb weaver *Metepeira grinnelli*, with the difference that lizards most likely depress spider numbers through intraguild predation rather than interference competition. Pacala & Roughgarden (1964) speculate that lizards prey upon juvenile spiders, which apparently place their webs lower in the vegetation than adults (no data are presented on web heights).

In this system spiders would have a substantial impact upon insect populations if lizards were absent, and might compete for prey. Predation by lizards may keep spider populations below competitive

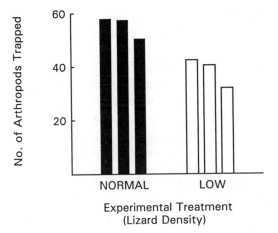

Fig. 8.1 Indirect effect of lizard abundance on numbers of arthropods trapped in the forest understorey. Number of arthropods > 1 mm caught over three days on 10 sticky traps suspended at heights of 1–2 m throughout each plot. Lizards were removed from six 12 × 12 m fenced plots and reintroduced into three plots at numbers approximating NORMAL density. No lizards were introduced into the LOW lizard-density plots. After six months lizard densities in NORMAL plots were two to three times higher than in LOW enclosures. Numbers of aerial arthropods were significantly higher in plots with more lizards [$F = 18$ (approx.), one-way ANOVA; $p < 0.05$, d.f. = 4; F is approximate because values have been taken from a figure in the original publication). (After Pacala & Roughgarden 1984.)

densities. Pacala & Roughgarden did not collect data on spider size or fecundity, so no inferences can be made about possible exploitative competition under increased spider densities. In the leaf litter, lizards prey heavily upon Diptera, which may be important prey for younger spiders. Thus the potential for exploitative competition between lizards and spiders also exists. The negative effect of lizards on spider density in the ground layer could have resulted at least partly from competition in addition to intraguild predation.

Pacala & Roughgarden's experiment was a pulse manipulation with respect to the alteration of lizard densities. Lizards were removed from fenced areas and then re-stocked in one set of pens, and were not introduced again. Immigrating or overlooked lizards were not removed from the 'no-lizard' treatment during the experiment. Although a pulse experiment, the perturbation of lizard densities persisted adequately, and the experiment was continued long enough, to reveal possible indirect interactions. Possible mechanisms of interaction between the species were not examined directly; the investigators interpreted the system's response to the manipulation of lizard numbers on the basis of natural history that was not collected as part of the study.

In contrast to the experiment of Pacala & Roughgarden, the study of Spiller & Schoener (1988, 1990a) on a similar community was a long-term press perturbation in which system attributes in addition to population densities of prey and consumers were monitored. Spiller & Schoener collected data directly relevant to separating effects of predation and competition in a long-term lizard-removal experiment on an island in the Bahamas. The entire experiment ran three years, but the first set of analyses was published half-way through the study (Spiller & Schoener 1988). The first 18 months of removing lizards produced a three-fold increase in the density of orb-weaving spiders, primarily *Metepeira datona* (experimental design and effects upon spiders are covered in Chapter 5). In contrast to what Pacala & Roughgarden found, Spiller & Schoener discovered that aerial insects were more abundant in the lizard-removal plots. Numbers were 20%, and biomass 50%, higher in the plots from which lizards had been continuously removed. Both differences are statistically significant [$p < 0.05$ and $p < 0.01$, respectively; family confidence levels for pairwise comparisons within the overall ANOVA (Tukey method)]. Spiller & Schoener propose two explanations for why their experiment did not reveal the type of indirect effect of lizards on insect abundance observed by Pacala & Roughgarden: (1) release from lizard predation did not produce as large an increase in spiders as Pacala & Roughgarden observed ($3\times$ versus $20\times -30\times$, respectively); and (2) differences between the two studies in vegetation heights and heights of insect traps might have influenced the ability to separate effects of lizards and spiders on aerial insect numbers.

Spiller & Schoener (1988) found evidence of competition between lizards and spiders, though the pattern was complicated. The mean percentage of *Metepeira* feeding was higher in the lizard-removal enclosures on three sampling dates ($21.5 \pm 2.9\%$, $4.3 \pm 0.2\%$ and $4.2 \pm 1.2\%$ in the experimental plots versus 0% on all dates for the control plots) − clear evidence, in conjunction with elevated prey densities on sticky traps, of exploitative competition between lizards and *Metepeira*. Feeding rates of *Metepeira* in the removal plots were also higher than in non-fenced controls, even though insect abundance was higher in the open plots. This discrepancy led to the hypothesis that an additional interaction between lizards and spiders was occurring: lizards possibly exclude *Metepeira* from patches of high prey abundance, either by predation or because the spiders tend not to forage in areas where lizards are active ('spatial shift hypothesis'). In order to test this hypothesis and also to elucidate other possible mechanisms underlying the negative effect of lizards on spiders, Spiller & Schoener continued removing lizards from the experimental plots and increased the number of parameters measured (Spiller & Schoener 1990a).

They continued the removal experiment for another 18 months, visiting the enclosures every three months instead of monthly as before. Two months before termination of the experiment, two control plots were re-stocked with lizards because densities had fallen below those in unenclosed control plots. At each census the abundance of aerial arthropods within the enclosures was sampled with sticky traps, and lizards found in the removal plots were killed and preserved in order to examine stomach contents. The experiment terminated in July 1988. During this month the investigators sampled insect abundances on a finer scale than previously, placing sticky traps in two types of locations: one where *Metepeira* webs were most dense, and the other where *Metepeira* was least common. Before collecting *Metepeira* and their egg sacs, Spiller & Schoener observed feeding rates of spiders in marked webs for eight days. Many response variables were measured at several times, some over a period of three years (encompassing the period discussed in the 1988 paper). Data that had been collected repeatedly were pooled before being analyzed statistically, thereby preventing problems of temporal pseudoreplication.

At the end of the experiment the difference in abundance of *Metepeira* between lizard–removal and control plots was still three-fold. The pattern was similar for all species of web spiders combined, and for juvenile and adult *Metepeira* considered separately (Fig. 8.2). The extent to which spider densities had increased in the lizard-removal plots was similar to the response observed 18 months into the experiment, and the fraction of the total number of web-building spiders represented by *Metepeira* (c. 45%) was similar to the average for the first 18 months of the study (c. 65%; calculated from Fig. 3 in Spiller & Schoener 1988).

Metepeira captured more prey in the plots from which lizards had been removed (Table 8.2), which translated into effects on growth and fecundity (Fig. 8.3). This effect, determined at the end of the experiment, was consistent with the overall influence of lizards on arthropod abundance during the three years of the study (Table 8.3). At the end of the study the investigators also obtained direct experimental proof that food was a limited resource for *Metepeira* in this habitat: supplementing

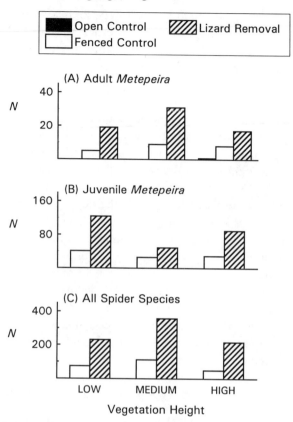

Fig. 8.2 Abundances (no. per plot) of web-building spiders after three years of removing lizards from enclosed plots. Plots are arranged by block based upon vegetation height. Numbers of adult *Metepeira* in open (unfenced) control plots are also presented (values are zero for first two blocks). Data for other categories of spiders in open controls were not reported. Differences in spider numbers between fenced-control and lizard-removal plots are statistically significant ($p = 0.025$, 0.028 and 0.008 for (A) adult *Metepeira*, (B) juvenile *Metepeira* and (C) all spider species, respectively; randomized block ANOVA, d.f. $= 1,2$; one-tailed criterion). (After Spiller & Schoener 1990a.)

the prey of adult females along a road near the enclosures more than doubled their fecundity. Thus the pattern of results for the entire three years of experimentation provides consistent evidence that exploitative competition for prey with lizards affects *Metepeira*. Lizards had a negative impact upon the feeding rate of *Metepeira* even though the overlap in lizard and spider diets was moderate (0.53 for taxonomic category, 0.63 for prey size; index of Schoener 1968). It appears that

Table 8.2. *Effect of lizard density on* Metepeira's *rate of prey capture. Replicate plots were divided into blocks based upon average height of the vegetation. The two size categories of prey (≤4mm and >4 mm) from Table 5 of Spiller & Schoener (1990a) have been combined and the data have been analyzed by univariate ANOVA (randomized block design) instead of by MANOVA as was done in the original publication. Simple ANOVA was used because the correlation between numbers in the two prey categories is weak (r=0.54, d.f.=4, n.s.)*

Treatment	Block (vegetation height)	Number of spiders observed	Average biomass of prey ($g \times 10^{-6}$) consumed/spider/ day
Lizards removed	Low	15	178.0
	Medium	27	90.6
	High	16	120.4
Control	Low	6	70.2
	Medium	5	29.2
	High	8	49.4

ANOVA

Source	d.f.	Mean square	F	p
Blocks	2	2093	6.95	0.126
Treatment – Lizards removed	1	9616	31.91	0.015[a]
Error	2	301		

Note:
[a] One-tailed test.
Source: Spiller & Schoener 1990a.

competition occurs because both lizards and *M. datona* are fond of the same common beetle species.

Removing lizards had no effect upon the manner in which *Metepeira* foraged for prey. At the end of the experiment, the height and area of webs did not differ between experimental and control plots, nor did the presence of lizards force spiders to forage in patches where aerial prey was less abundant. Thus there was no evidence that behavioral interference from lizards affected this orb weaver; the 'spatial shift hypothesis' was not supported.

Early in their experiment Spiller & Schoener found that survival of marked *Metepeira* was lower in plots with lizards (Chap. 5). Spiders

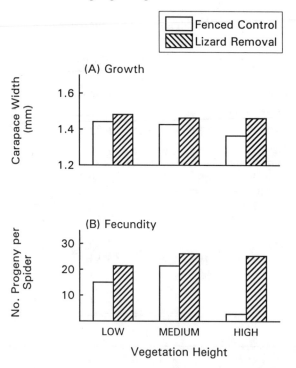

Fig. 8.3 Effect of reducing lizard densities on (A) growth and (B) fecundity of the orb weaver *Metepeira*. Because carapace width and fecundity are correlated ($r = 0.87$, $p < 0.05$, d.f. = 4), the investigators analyzed the data with multivariate ANOVA (randomized block design). The effect of lizards in the MANOVA is significant ($p = 0.031$, d.f. = 2,1). Probabilities of F's for separate univariate ANOVAs of carapace width and fecundity are 0.047 and 0.093, respectively (d.f. = 1,2, one-tailed test). (After Spiller & Schoener 1990a.)

comprised 10% of lizard diets in the enclosures. Thus predation by lizards, in addition to decreased growth and fecundity due to competition with lizards, depresses densities of *Metepeira*. Spiller & Schoener speculate that predation may be the more important interaction, but point out that to assess the relative contribution of the two processes in this system would require additional study. Although direct experimental evidence relates only to effects of lizards on spiders, information on relative densities, sizes and diets of both taxa suggests that the negative interactions of intraguild predation and interspecific competition are primarily one-way, with orb-weaver populations being depressed by lizards. Intraguild predation between the orb weavers of this community appears to be negligible, which is probably true for most assemblages of

Table 8.3. *Effect of removing lizards for three years on the cumulative number of arthropods captured in sticky traps. Arthropods were sampled on 21 dates with four sticky traps/enclosure. Replicate plots were divided into blocks based upon average height of the vegetation. The two size categories of prey (≤4 mm and >4 mm) from Table 10 of Spiller & Schoener (1990a) have been combined and the data have been analyzed by univariate ANOVA (randomized block design) instead of by MANOVA as was done in the original publication. Simple ANOVA was used because the correlation between numbers in the two prey categories is weak (r=0.46, d.f.=4, n.s.)*

Treatment	Block (vegetation height)	Arthropod biomass $(g \times 10^{-3})$
Lizards removed	Low	148.7
	Medium	146.4
	High	180.2
Control	Low	110.3
	Medium	104.3
	High	128.2

ANOVA

Source	d.f.	Mean square	F	p
Blocks	2	486.9	19.7	0.048
Treatment – Lizards removed	1	2930.5	118.4	0.004[a]
Error	2	24.8		

Note:
[a] One-tailed test.
Source: Spiller & Schoener 1990a.

web spiders. The relatively low densities of spiders also suggests that exploitative competition for prey among spiders is not important in this system.

The study of Spiller & Schoener is remarkable for its duration, and for its attention to uncovering mechanisms underlying the observed responses of the system to the sustained perturbation. Few press experiments of this scope have been conducted. The benefits of this approach are clear – a non-manipulative study would not have been able to establish the extent of the negative effect of lizards upon spiders, and a perturbation experiment without supporting studies would not have yielded the understanding of mechanism their research has provided.

Their experiment also provides a lesson in the difficulties inherent in attempting to draw unambiguous conclusions from a deceptively simple and seemingly straightforward field experiment. The problem lies in what are innocuously termed 'fence effects.' They always pose potential problems of interpretation, even in experiments similar to this one, in which fenced areas were large (84 m²). Fortunately, the experimental design of Spiller & Schoener (1988, 1990a) included open control plots, making it possible to gauge the impact of fencing. Pacala & Roughgarden (1984) had no control for fencing in their design, so that one will always wonder if the dramatic response of spiders to lowered lizard densities was due to the impact of fencing on species interactions in the system. This is not a trivial point, as Spiller & Schoener discovered significant fence effects:

(1) Numbers and biomass of aerial arthropods were approximately 20 and 50% higher, respectively, in the unenclosed control plots than in the fenced control plots. Thus comparing spider parameters in fenced lizard-removal plots to those in fenced control plots may reveal an effect of lizards that is at least partly artificial, i.e. that was induced by lower prey abundance caused by the fencing. Spiller & Schoener (1988, 1990a) recognize this possibility, but judge it to be unimportant because temporal fluctuations in prey numbers are greater than the decrease induced by fencing the plots. The net effect may be to over-estimate the importance of competition versus predation.

(2) Spiller & Schoener point out that constructing the plots disrupted the spider populations so much that adult *Metepeira* had to be added to the plots at the start of the experiment. This does not present a problem in interpretation, though, because the pattern of population density in the open controls resembled that of the fenced control. Numbers of *Metepeira* were also unusually low in the open plots at the start of the experiment. Pacala & Roughgarden (1984) added lizards to their plots to bring them to what they judged to be normal densities, which introduces an element of artificiality into their design.

(3) Near the end of the experiment lizard numbers dropped in two of the fenced control plots, leading the investigators to re-stock the plots 'to match natural densities' (Spiller & Schoener 1990a). No data are presented, but the drop is large enough to cause some concern. Might the fence have been a factor in this decline?

(4) At the end of the experiment *Metepeira* was effectively absent from the open control plots (Fig. 8.2A), forcing Spiller & Schoener to omit them from their analyses. They attribute this decline to a possibly 'random divergence in spider abundance.' An alternative interpretation is more consistent with the philosophy of their experimental design: the fencing, in addition to affecting numbers of aerial insects, had a significant impact upon spider populations. The investigators speculate that the enclosures may increase spider densities by ameliorating effects of wind.

Ambiguities introduced by fence effects are not the result of a flawed experimental design, but are an annoying example of the complications inherent in the manipulative approach. In fact, it was the solid design of the experiment (i.e. the inclusion of replicate open controls) that made it possible to evaluate the consequences of enclosing the plots. Another reason the fence effects appeared is the length of the experiment, but the long duration of the study is also one of its virtues. On balance, the approach taken by Spiller & Schoener to this system has been highly fruitful, and provides an excellent example of one productive way to unravel the ecological web to which spiders belong.

Wolf spiders and mantids

Hurd & Eisenberg (1990) investigated direct interactions between the wolf spider *Lycosa rabida* and the mantid *Tenodera sinensis*, and also examined the possible impact of such interactions upon the pressure exerted by these generalist predators upon the arthropod community. The experiment was conducted in a horse pasture, in a fenced 2 ha portion in which the vegetation had grown high enough to provide suitable habitat for *Tenodera*. Using a completely randomized design, Hurd & Eisenberg established the following four treatments (five replicates each) in field cages each covering 1 m^2 of pasture: addition of 8 mantids, addition of 10 lycosids, addition of mantids and lycosids (8 + 10), and a control – neither mantids nor lycosids added. Mantids had not colonized this pasture, but did occur in other, less disturbed fields at densities comparable to the number introduced into the cages. Adding wolf spiders produced an initial density within the enclosures approximately 4× normal. Thus the control enclosures had no mantids but housed *Lycosa* at densities typical of the open habitat. After 10 days arthropods in the 20 enclosures were collected with a suction sampler and by hand searching. The experimental design reveals whether or not interactions within this predator guild influence its overall impact upon the community. A significant interaction term in a two-way ANOVA would be a good indication [but not necessarily final proof – cf. discussion in Hurd & Eisenberg (1990)] that such intraguild interactions influence the survival of other arthropods in the community.

Adding mantids had no statistically significant effect upon the total number of other arthropods (non-lycosids and non-mantids) in the community, but did significantly reduce numbers in the two largest size classes [$p < 0.05$, table-wide criterion based upon the sequential Bonferroni procedure (Rice 1989)]. Total arthropod biomass, as well as biomass within the two largest size classes, was also significantly lower in the presence of mantids. In contrast, adding lycosids had no impact upon numbers or biomass of other arthropods. The mantid × lycosid interaction term was non-significant for all comparisons.

Whether or not wolf spiders have an impact upon the natural community outside the cages cannot be determined from this experiment for two reasons: (1) lycosid numbers declined to control levels in the enclosures to which they had been added, and (2) there was no treatment in which wolf spider densities had been reduced below natural values. Densities of lycosids at the end of the experiment did not differ significantly between any of the treatments ($p(F) > 0.25$, d.f. = 3,16), though the mean number ± s.e. in the lycosid-addition treatment was slightly higher than the control (7.2 ± 1.0 and 3.4 ± 1.5, respectively). Cannibalism is the likely explanation for the rapid decline to control levels. Some cannibalism probably occurred among mantids, but their survival rate over the 10 days of the experiment was ≈60% (same in both mantid-addition treatments). There was no evidence that exploitative competition affected either mantids or lycosids – average weights did not differ among treatments. Thus all effects upon survival of these predators must have resulted from cannibalism or IGP.

Focusing on particular components of the arthropod fauna reveals some intriguing patterns. Mantids had a major effect upon densities of grasshoppers, but not crickets (Fig. 8.4A, Table 8.4). This result agrees with the natural history of predator and prey (Hurd & Eisenberg 1990): mantids and grasshoppers are most active in the upper vegetation, whereas crickets forage primarily on the ground. Wolf spiders also forage on the ground and are predators of crickets, yet adding lycosids enhanced the density of crickets (Fig. 8.4B). The investigators do not provide an F-statistic for this main effect, because the interaction term was statistically significant. Nevertheless, as they imply in their discussion of the results, there was a believable main effect: the significant

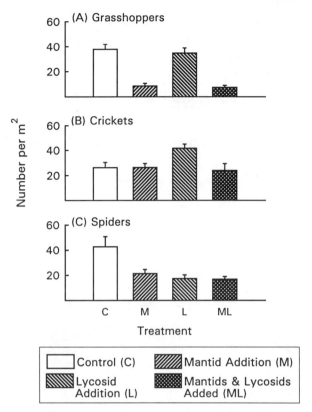

Fig. 8.4 Responses of selected components of the arthropod community in a pasture to additions of the mantid *Tenodera sinensis* and the wolf spider *Lycosa rabida* to 1 m² field enclosures. Means ± s.e. (*n* = 5). (A) grasshoppers, (B) crickets, (C) spiders ≤ 8 mm. (After Hurd & Eisenberg 1990.)

interaction term reflects the disappearance of the enhancement effect when mantids were added. Hurd & Eisenberg speculate that increased interference among wolf spiders at high densities reduced their effectiveness as cricket predators, and that mantids counteracted this effect in the mixed treatment by reducing lycosid numbers. The fact that the effect of mantids on *Lycosa* numbers was not statistically significant argues against this explanation, but the interaction between mantids and wolf spiders could have occurred only at the beginning of the experiment, before cannibalism lowered lycosid densities. Elevated activity of wolf spiders also may have induced crickets to escape to higher vegetation, which would have improved their short-term survival, but only in the absence of *Tenodera*.

Another hypothesis to explain enhanced cricket densities is facilitation

Table 8.4. *Results of two-way ANOVAs examining the effects on numbers of selected arthropods of adding the mantid* Tenodera sinensis *and the wolf spider* Lycosa rabida *to field enclosures.* $M =$ *mantid main effect;* $L =$ *lycosid main effect;* $M \times L =$ *interaction*

Group	Source	d.f.	Mean square	F^a	p^b
Grasshoppers	M	1	3808.8	73.11	< 0.0001*
	L	1	33.8	0.65	0.4408
	M × L	1	1.8	0.04	0.8568
	Error	16	52.1		
Crickets	M	1	378.4	. . . .	. . . .
	L	1	211.2	. . . .	. . . .
	M × L	1	396.0	10.50	0.0051*
	Error	16	37.7		
Spiders[c]	M	1	583.2	7.66	0.0137
	L	1	1065.8	14.00	0.0018*
	M × L	1	520.2	6.83	0.0188
	Error	16	76.1		

Notes:
[a] *F*-values for main effects not calculated if interaction term is significant.
[b] (*) = significant at a table-wide α of 0.05 (sequential Bonferroni procedure).
[c] Spiders ⩽ 8 mm; these spiders are smaller than the introduced *L. rabida*.
Source: Hurd & Eisenberg 1990.

caused by *Lycosa*'s consuming other cricket predators. Densities of small spiders (⩽ 8 mm) were significantly lower in the lycosid-addition enclosures (Fig. 8.4C, Table 8.4). However, numbers were also lower in the other two predator-addition treatments, though the effect was judged not significant because table-wide α values were commendably used. Nevertheless, the pattern of densities of small spiders does not convincingly explain the pattern in cricket abundances across treatments.

Hurd & Eisenberg's study is a short-term pulse experiment that was designed to uncover short-term direct and indirect effects, i.e. those resulting from immediate consequences of IGP. The investigators terminated the experiment after only 10 days in order to eliminate the possibility that exploitative competition for prey would influence survival of *Lycosa* or *Tenodera*. Even over this short period the 'pulse perturbation' of lycosid density had effectively disappeared, revealing information about interactions among lycosids, but weakening conclusions about interactions with mantids and the rest of the arthropod community. A more difficult but more revealing pulse perturbation would have been to remove *Lycosa* by pitfall trapping. Larger experimental units would be required with this approach, since the natural density of *Lycosa* in this habitat was 3–4/m². These are easy suggestions for a critic to make in hindsight. The study of Hurd & Eisenberg is a well-designed experiment that was analyzed correctly and interpreted cautiously but imaginatively.

Other examples of IGP

Spiders share their trophic levels with a variety of other predatory groups. For example, ants are important generalist predators in terrestrial systems. Wandering spiders and ants prey upon each other (Pętal & Breymeyer 1969, Pętal 1978), opening up the possibility of a mix of competitive and predatory effects between them. Breymeyer (1966; cited by van der Aart & de Wit 1971) interpreted patterns of pitfall catches as evidence of possible competitive interactions between ants and spiders. No experimental evidence exists for competition between ants and spiders. Van der Aart & de Wit (1971) reanalyzed Breymeyer's data and found no evidence of a negative relationship between ant and spider abundances, a result consistent with the data they collected. Because the scope of these non-experimental studies is limited, the question of whether ants and wandering spiders affect each other's population dynamics remains unanswered.

Centipedes and spiders likely interact. Albert (1976) documents a negative correlation between abundances of spiders and centipedes in a beech forest, a pattern consistent with both exploitative competition and IGP. In discussing the results of their spider-removal experiment in a beech-litter community, Clarke & Grant (1968) point out that possible consumption of Collembola by centipedes complicates the interpretation of a simple spider-removal experiment. Because spiders prey upon both Collembola and centipedes, Collembola populations may not benefit from a reduction in spider numbers if the lower density of spiders increases the predation pressure on Collembola from their centipede enemies. One needs to know the relative predation rates of all the possible pathways in this web, including the effect of centipede predation upon spiders. A complete understanding would also require knowledge of how predation rates change as a function of changing predator size during development. Younger spider instars may be vulnerable to predation by centipedes, but those that escape may prey upon their former enemies once they become large enough.

Polis *et al.* (1989) argue that IGP is more widespread than is generally realized. According to their assessment, the ubiquity of the process 'questions the generality of many proposed food web patterns.' Intra-guild predation, in addition to exploitative competition, can explain patterns of guild structure. Polis *et al.* (1989) argue that field experiments are needed to separate the impacts of IGP and exploitative competition; it would be worthwhile to continue pursuing this question with respect to spiders and other generalist predators with which they potentially interact.

Complications of IGP create fascinating wrinkles in the ecological web, but are by no means the only indirect effects involving spiders. For example, do changes in densities of spiders alter herbivore damage to plants? Schoener (1988) established that the amount of damage to buttonwood trees among islands in the Bahamas was negatively corre-lated with lizard density. He searched for such a correlation between the abundance of orb weavers and insect damage to buttonwood, but found none, even when only islands without lizards were compared. It appears that orb weavers do not depress populations of herbivorous insects to levels where buttonwood damage is lessened, even when spider popula-tions are released from the pressure of lizard predation. In this system lizards have the major effect, but in others predation pressure from spiders can be sufficient to protect plants from some herbivore damage. An excellent example is the experimental demonstration by Riechert & Bishop (1990; Chap. 6) that spiders decrease plant damage from pest insects in a vegetable-garden community. Below I review the results of other studies designed to reveal whether primary producers experience consequences of spider predation.

Cascading effects of spiders in grazing food chains

What influence might spiders exert upon plants by controlling densities of herbivorous insects? Fretwell (1987) proposed that the approach to food-chain dynamics pioneered by Hairston *et al.* (1960) and subse-quently developed by other ecologists be considered a central theory of ecology. In selecting an example of an experimental test of one tenet of this theory he proposed plucking the spider's ecological web: 'If elimination of spiders from an arid shrubby desert causes defoliation of the desert shrubs, and their eventual replacement by more rapidly reproducing herbs and grasses, much of the development of the theory ... (of food chain dynamics) ... since the mid-1970s must seriously be questioned, especially as to generality.' Fretwell argues that the response of a system to removing predators such as spiders depends upon the rate of primary production, an interplay between donor control and top-down control of population dynamics in the ecological web. Other predators would of course be appropriate foci for experimental probes of cascading effects in terrestrial food webs, but it is heartening when non-arachnologists select the spider as an exemplar.

Top-down effects: spiders benefiting plants

Louda (1982) uncovered evidence that the green lynx spider *Peucetia viridans* (Oxyopidae) benefits a small shrub on which it forages for prey. The shrub grows in disturbed coastal scrub communities of western North America. On a single day at the end of the growing season Louda sampled the tallest central flowering branch of 20 plants – 10 with, and 10 without, *Peucetia*. At this time the adult females are guarding their egg sacs. The number of damaged flower heads was lower on branches with the oxyopid, but the number of pollinated flowers per flower head was also lower (Table 8.5). Louda interpreted this pattern as resulting from predation by *Peceutia* on seed predators and pollinating insects. He calculated that the net effect of the spider was positive: numbers of viable seeds released per branch were 18% higher on branches with *Peceutia* (Table 8.5). As Louda points out, the overall effect of this higher-order interaction is sensitive to when spiders colonize the plants and the timing of flowering. A slight shift in either phenology could change the spider's presence from a benefit to a detriment.

A single sample of the growing season cannot provide convincing proof of the postulated impact of the green lynx spider on its host shrub's reproductive rate. However, the correlations uncovered by Louda lead to an intriguing hypothesis, one that merits an experimental test. It would be fascinating to devise and implement a non-intrusive procedure for excluding *Peceutia* from some plants, but not others, and then monitor reproductive output of the plants.

Although Schoener (1988) found no indirect evidence that spiders benefit buttonwood, Spiller & Schoener (1990b) obtained experimental evidence that spider predation has the potential to indirectly benefit sea grape by preying upon gall midges that damage its leaves. Spiller & Schoener (1990b) report that removing lizards in their long-term field experiment (Spiller & Schoener 1988, 1990a) reduced leaf damage caused by gall midges ($p(F) = 0.043$, d.f. $= 1,4$). Apparently the tripling of spider numbers that occurred when lizard densities were lowered (cf.

Table 8.5. *Indirect evidence that the green lynx spider,* Peucetia viridans, *affects the reproduction of the shrub on which it forages for prey. Damage to flower heads, number of pollinated flowers per head, and net number of viable seeds produced, on the tallest branch of 10 shrubs (*Haplopappus venetus*) on which the spider was present, and 10 shrubs on which it was absent. The statistical test for an effect of the spider on total number of pollinated flowers/head may be sacrificially pseudoreplicated, because sub-samples within the bush on each plant were used to calculate the standard error. It is not clear whether or not the average value per branch was used to calculate the Mann–Whitney U-statistic. None of the other statistical comparisons is pseudoreplicated, because values per branch were compared*

	Without *Peucetia*		With *Peucetia*		
	n	Mean ± s.e.	*n*	Mean ± s.e.	p^a
Total no. of flower heads[b]	10	78.9 ± 9.9	10	81.8 ± 8.1	n.s.
No. of damaged flower heads[b]	10	67.9 ± 10.2	10	45.8 ± 6.5	*
No. of pollinated flowers/head:					
Heads with no insect damage	22	5.4 ± 0.7	262	3.6 ± 0.5	*
Heads with insect damage	515	5.4 ± 0.5	277	3.9 ± 0.4	*
No. of viable seeds produced[b]	10	243 ± 10	10	286 ± 12	*

Notes:
[a] Probability of Mann–Whitney U-statistic; $* = p \leqslant 0.05$.
[b] No./branch.
Source: Louda 1982.

Chap. 5) led to a reduction in midge densities. The magnitude of the spider impact on midges is unknown for the situation when lizards are present at normal densities.

Andrzejewska *et al.* (1967) designed a field experiment to determine the impact of an orb-weaving spider on plant production. They stocked seven field enclosures in a Polish meadow with grasshoppers and different numbers of a common orb-weaving spider, *Araneus quadratus*. At the end of the study they found a negative correlation between the average number of spiders present in each cage during the course of the experiment, and the fraction of plant matter estimated to have been destroyed by grasshoppers (Fig. 8.5). Unfortunately, this result may have minimal bearing on the impact of spiders on vegetation in a natural meadow. Treatments and controls were not established and censused

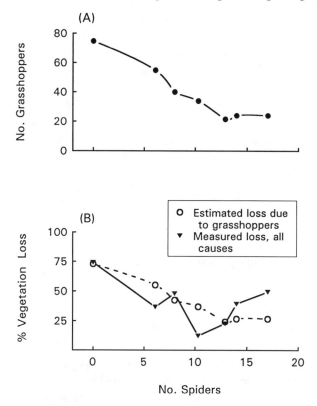

Fig. 8.5   Relationship between density of the spider *Araneus quadratus* in experimental enclosures in a Polish meadow and (A) density of grasshoppers, and (B) measured loss of vegetation and estimated loss due to consumption by grasshoppers. Spider and grasshopper densities are the mean numbers in each of seven cages (0.64 m²) over the course of the six-week experiment. The negative correlation between average spider density and average grasshopper numbers is highly significant ($r = -0.96$ $p < 0.001$, d.f. $= 5$). Estimates of the amount of vegetation consumed by grasshoppers were obtained by multiplying grasshopper densities by the same constant. There is no statistically significant correlation between spider density and the measured rate at which grass disappeared. (From data in Andrzejewska et al. 1967.)

contemporaneously, and the amount of plant material consumed by grasshoppers was estimated by multiplying the number of grasshoppers in each treatment by the same factor. No correlation was found between the number of spiders in a cage and the measured loss of grass during the experiment (Fig. 8.5B). The major shortcoming, however, lies in the spatial scale and densities of animals selected for experimentation. Field cages were 1 m high and 0.64 m² in area, which would enclose an average of seven grasshoppers according to censuses of the open meadow (Andrzejewska *et al.* 1967). Yet the investigators added 200 grasshoppers to each cage, with no justification of the decision to employ a density over an order of magnitude greater than natural. This design is so artificial that extrapolation of the results to natural conditions is practically impossible. Larger cages with many fewer grasshoppers should have been used. If this were not feasible because of logistical limitations, it would have been more productive to have devoted the saved effort to gathering detailed information on feeding rates of spiders and grasshoppers on open meadows, at natural densities under undisturbed conditions. The study of Andrzejewska *et al.* (1967) is an example of how a well-intentioned but inadequately designed field experiment may yield clear results that are less illuminating than the more ambiguous patterns that an intensive, non-experimental research program might have furnished.

Bottom-up effects: responses of spiders to enhanced plant productivity

Experimentalists have also searched for indirect effects propagated in the opposite direction, with spiders on the receiving end. Perturbing the bottom of the food chain by fertilizing plants produces responses by spiders and other top predators.

Salt-marsh communities

Experiments in salt marshes demonstrate that spider populations respond to enhanced productivity of plants induced by fertilization, and they provide indirect evidence that spiders limit insect numbers. Results of a fertilization experiment by Vince *et al.* (1981) suggest that spiders in salt marshes limit population densities of the planthopper *Prokelisia marginata*. The investigators boosted the herbivore trophic level by fertilizing the grasses in open, 300 m² plots. Fertilization of the marsh improved the quality of the food of herbivorous insects, leading to an increase in the density of the total herbivore trophic level. Numbers of the planthopper *P. marginata*, however, were not higher on the fertilized plots. Densities of *P. marginata* exhibited a negative correlation with spider numbers across treatments (Fig. 8.6), which suggests that increased spider predation prevented planthopper populations from responding to fertilization, in contrast to other phytophagous insects in

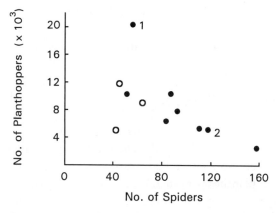

Fig. 8.6 Negative relationship, across treatments, between total number of spiders and abundance of the planthopper *Prokelisia marginata* in a salt-marsh fertilization experiment. Abundances are expressed as seasonal totals. Open circles – control plots; closed circles – fertilized plots. Four different fertilization treatments were used, but are not distinguished here. Planthopper densities were highest in a fertilized plot (No. 1) in which spider numbers did not respond as rapidly to elevated densities of other herbivores as occurred in the other replicate of the same fertilizer treatment (No. 2). (After Vince *et al.* 1981.)

this system. Spider populations apparently exhibited a numerical response to increased density of the entire herbivore assemblage, leading to an increase in predation pressure on *P. marginata* that was sufficient to prevent the expected increase in planthopper density. Vince *et al.* conclude that *P. marginata*'s size and its location in the grass canopy expose it more than other salt-marsh herbivores to spider predation.

Döbel (1987) also obtained results from fertilization experiments that suggest spiders can limit *Prokelisia* spp. populations in salt marshes. He conducted his experiments in pure stands of *Spartina alterniflora*. Over 90% of all Homoptera in the marsh belonged to two species of planthopper, *P. marginata* and *P. dolus*, which accounted for over half of the prey in the diets of the two most common lycosids. Twenty to 60% of the spiders sampled over the season were wolf spiders. Döbel studied lycosid–planthopper interactions by manipulating each component indirectly. Two 100 m² plots were fertilized in three years in order to increase planthopper populations; two untreated plots were controls. In the third year Döbel expanded the design to include two plots in which thatch was removed in order to lower wolf spider densities. Fertilizing the plots increased the biomass of *S. alterniflora* in each year, but responses of the planthoppers and spiders varied between years. In the first year *Prokelisia* nymphs and wolf spiders were higher in the fertilized plots. In the second year's experiment (actually conducted two seasons later), planthopper densities were not higher in the fertilized plots, though spider densities were significantly greater. At season's end adult planthoppers were lower in the fertilized plots compared with controls. In the next season fertilization produced responses by planthoppers and spiders similar to those of the first year, although overall density of planthoppers was lower, and densities of spiders were higher, than in the first year. Lycosids were less dense in the de-thatched (raked) plots, but planthopper populations were not elevated compared to control plots. Absence of a response by planthoppers in the de-thatched treatment suggests that lycosids may not limit planthopper numbers; however, the reduced amount of thatch may have adversely affected the planthoppers. There was no control for the manipulation of raking the plots. The correlations between planthopper and spider densities for the three years are summarized in Fig. 8.7.

Döbel argues that the results can be interpreted as evidence that lycosids play a role in limiting planthopper populations, though he recognizes that the pattern is clearly complicated. In the first and third years lycosids showed a numerical response to increases in *Prokelisia* numbers; in the middle year the negative correlation between spider and planthopper densities suggests that spiders can limit planthopper densities under certain conditions. Döbel explains the differences between years as resulting from a threshold density of *Prokelisia* above which lycosids are unable to limit population growth of the planthopper. In the middle year, planthopper densities were lower at the start of the season than in the other two years, but initial wolf spider densities were not lower. Döbel speculates that in this year spiders were able to limit growth of the planthoppers because the ratio of spiders to planthoppers was higher at the beginning of the season than in other years. These differences in relative abundances of spiders and planthoppers may at least partially result from time lags inherent in the predator–prey interaction.

The impact of spiders on *Prokelisia* populations in the salt marsh studied by Vince *et al.* (1981) appears to be greater than in Döbel's marsh. Time lags in the interaction between spiders and *Prokelisia* may differ between the two communities because of differences in community structure. Also, species diversity of the herbivore community may have been lower in Döbel's system. In a simpler food web there would be less opportunity for the spiders to show differential response to, and effects

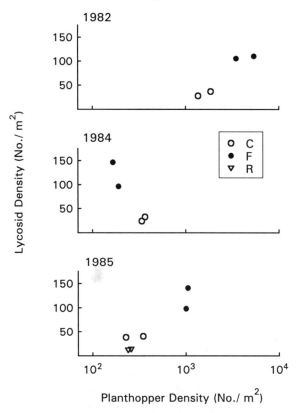

Fig. 8.7 Effects of fertilization and removal of thatch on abundances of planthopper nymphs (*Prokelisia*) and lycosids in a salt marsh in three different years. C = unfertilized control plots; F = fertilized plots; R = thatch removed by raking, not fertilized. (After Döbel 1987.)

upon, different types of herbivores. The potential number and magnitude of indirect effects would be fewer in the less-complex community. Temporal and spatial heterogeneity, both between and within the two study areas, should make one cautious in generalizing too much without additional experimentation. It turns out that Vince *et al.* also uncovered differences between years in the response of spiders to elevated numbers of *Prokelisia* in one of their plots (Fig. 8.8).

Perturbing the base of the food web in salt marshes has yielded fascinating information. The primary goal of such manipulations was not to reveal the impact of spiders on their prey, yet such nutrient enhancements have yielded insights because of the apparent chain-reaction that begins when spider populations react to increases in prey

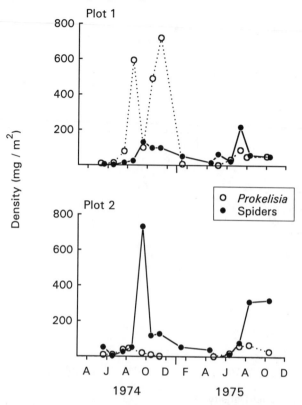

Fig. 8.8 Variation between plots in the response of spider populations to increases in densities of the planthopper *Prokelisia* caused by fertilization of a salt marsh. Plots identified as 1 and 2 are plots 1 and 2 in Fig. 8.6. (After Vince *et al.* 1981.)

numbers. Evidence obtained on spider limitation of insect numbers, though compelling, is still indirect because spiders were not manipulated directly. Nevertheless, manipulating the base of the food web in these marshes has provided worthwhile insights. These studies demonstrate how field experiments can provide a direct means to uncover unexpected interactions in complex ecological systems.

An agroecosystem

Ecologists have also examined the impact of fertilization upon spiders in agroecosystems (Kajak 1980b). A major experiment conducted by Polish ecologists involved fertilizing plots on a meadow over a eight-year period (Kajak 1978b, 1980a, 1981). A meadow that was mown regularly each year was divided into eighteen 1250 m² unfenced plots. Six plots were fertilized yearly with 280 kg NPK/ha, another six with 680 kg NPK/ha, and six were left as controls. Primary production increased on the fertilized plots, accompanied by a decrease in the number of plant species; almost three-quarters of the plant biomass on the fertilized plots was due to two species (Traczyk, Traczyk & Pasternak 1976; cited by Kajak 1981). Spider densities were sampled by extracting soil/litter cores, by examining vegetation within small sampling frames, and by setting pitfall traps. Kajak (1981) reports that fertilizing the plots affected neither the total density of spiders, nor the species diversity of spiders as measured by Simpson's Index and the Shannon–Weaver Index. However, regular fertilization of the plots apparently caused spider biomass to decline. This reduction resulted from a shift in the composition of the community from one dominated by wolf spiders to one in which the smaller linyphiids were more common. The change occurred even though spiders could move freely from untreated to treated areas.

Though this field experiment appears to have been well designed and adequately replicated, it is not certain that the effects judged by Kajak (1981) to be statistically significant reflect the effect of adding fertilizer. Three features lead one to question the firmness of the conclusions: (1) mean spider weights were not significantly different in all years of the experiment; (2) Kajak describes the number of samples taken within each plot every year and states that control and experimental plots were compared with t-tests, but she does not give the degrees of freedom. Thus, differences between control and fertilized plots in mean weight and numbers of individuals of lycosids or linyphiids may be statistically significant because of sacrificial pseudoreplication in the analysis of the data (Hurlbert 1984); (3) Given the uncertainty about the design of the statistical test employed, it would be helpful to know initial conditions of the plots in order to judge whether the differences between controls and treatments might reflect divergence over time due to fertilization. However, initial conditions, i.e. spider densities and sizes, were not determined in the plots the year preceding the first fertilizer application.

Kajak sampled spiders that prey heavily upon insects and other invertebrates of the ground layer; these spiders belong less to the grazing food web, and more to one that is based ultimately upon energy in plant detritus. Kajak (1981) speculates that the apparent change in family composition of the spider community is a reaction to a change in the abundance and/or species composition of saprophagous species. She also hypothesizes that a shift in composition of the spider community to smaller species, and less total biomass, might lessen predation pressure on saprovores, accelerating the rate of decomposition. This is a reasonable conjecture, one that deserves testing with field experiments.

Spiders in detritus–based food webs: possible effects on decomposition

Clarke & Grant's (1968; Chap. 6) removal experiment was designed to measure the impact of spider predation upon Collembola densities in a forest-floor community. A demonstrated effect would have prompted the question of whether spiders influence the rate at which leaves decompose, given the role of Collembola in decomposition. No one has yet attacked this question directly through field experimentation in forest-floor communities, but Kajak & Jakubczyk (1975, 1976, 1977) have attempted to determine whether spiders and other generalist arthropod predators influence rates of litter disappearance in a grassland.

Kajak & Jakubczyk used small exclosures to alter predator densities. They did not directly manipulate densities of spiders and other predators; instead, they altered the accessibility of mobile predators to 100 cm^2 × 15 cm soil/litter cores in 0.3 mm-mesh nylon bags, by inserting windows of different-sized mesh (1 mm or 6 mm) at the interface between litter and soil. They followed the rate of disappearance of 2 g of hay in each bag in order to measure the impact of their treatments upon rates of decomposition. In the first year's experiment Kajak & Jakubczyk (1975, 1976) used 50 bags of each of the three treatments; 25 of each were sampled after 16 days, the remainder after 50 days. At the beginning of the experiment, and on the two sampling dates, the investigators also sampled 30 similarly sized soil cores from the open meadow. At both sampling times bags that permitted immigration of predators had significantly fewer Collembola than the closed bags (Table 8.6). Kajak & Jakubczyk interpret this difference as primarily due to differences in predation by spiders and carabids, rather than increased emigration of Collembola from the more open bags, or differential alteration of microclimate by the windows of differing mesh size. The rate of litter disappearance was highest in the closed-bag treatment, even though moisture content of the litter was slightly lower than in the bag with the 6-mm-mesh window. This pattern is opposite to what an altered microclimate would have caused, and suggests an indirect effect of spiders on litter decomposition mediated through an interaction with

Table 8.6.*Collembola numbers and rates of litter decomposition in soil cores placed in closed bags and bags open to predators (6-mm-mesh windows) in a grassland. Results for bags with fine-meshed windows (1 mm) have been omitted*

			Treatment		
Exp.	Duration (days)	Open meadow[a]	Closed bag	Open bag	t[b,c]
Collembola (no./100 cm²)					
1973					
A	16	4.1	5.2 ± 0.5	3.3 ± 0.3	**
B	50	119	100 ± 13	38 ± 7	*
1974					
A	12		305 ± 13	210 ± 6	6.45**
B	16		80 ± 6	56 ± 2	3.96**
C	20		99 ± 3	87 ± 2	4.26**
D	25		64 ± 3	53 ± 2	3.35**
Litter disappearance (mg/g/d)					
1973					
A	16		1.43 ± 0.10	1.57 ± 0.18	n.s.
B	50		3.56 ± 0.69	2.94 ± 0.71	n.s.
1974					
A	12		6.53 ± 3.20	4.99 ± 1.66	0.56
B	16		1.42 ± 0.62	1.46 ± 0.65	−0.19
C	20		0.62 ± 0.18	0.50 ± 0.14	1.20
D	25		1.94 ± 1.66	1.02 ± 0.85	0.64

Notes:
[a] Areas outside the bags were sampled only in 1973. Values are means of 30 samples; no measure of sampling variation was reported.
[b] Only the probability levels, and not the t-statistic, were reported for 1973 data. Table-wide αs were not calculated, but the consistent pattern of statistical significance of the individual t-tests indicates that the treatment influenced Collembola densities.
[c] Probability levels: $* = p < 0.05$; $** = p < 0.01$.
Source: Kajak & Jakubczyk 1975, 1976, 1977.

Collembola populations. Differences between treatments in rates of litter disappearance, however, were not statistically significant. Several categories of microflora were also sampled, but no consistent pattern emerged.

The next year the investigators repeated the experiment and again obtained evidence that predators affect Collembola numbers (Kajak & Jakubczyk 1977). The design was similar, but consisted of four separate experimental runs of shorter duration (12–25 days). Collembola densities clearly were higher in the completely closed bags (Table 8.6). Disappearance rates of litter were generally less in the open bags (Table 8.6), leading the investigators to suggest that 'predators reduce the rate of decomposition processes' (Kajak & Jakubczyk 1977). As in the previous year, though, differences were not statistically significant.

Limiting access of predators to the bags had a consistent impact upon Collembola numbers. This result is not direct proof of the importance of predators because they were not manipulated directly. Kajak & Jakubczyk cite indirect evidence that predators lowered Collembola densities in the open bags. In both years the researchers placed pitfall traps in a series of bags with large-mesh windows in order to remove mobile predators that had immigrated into the bags; the traps captured mainly spiders and carabids. Numbers of Collembola in open bags with pitfall traps did not differ significantly from Collembola densities in closed bags, which suggests that the significant treatment effect resulted from altering accessibility of the soil cores to predators. Kajak & Jakubcyzyk also cite indirect evidence that spiders, and not other generalist predators, caused most of the reduction in Collembola numbers. By chance a few closed bags contained one or more spiders at the start of the first year's experiment; at the end of the study these enclosures tended to have slightly fewer Collembola than bags without spiders (87 versus 107 Collembola per enclosure, respectively; no estimate of sampling variation given). The number of bags that happened to contain spiders '. . . was too low for statistical analysis,' so the meaning of this apparent difference is unknown. Numbers of replicates (5,9) were adequate for statistical analysis, but none is given. No similar difference in Collembola numbers occurred between bags with and without carabid beetles.

The two years of experiments by Kajak & Jakubczyk produced clear evidence that the treatments affected Collembola densities, probably due to predation by spiders. There was no significant difference between treatments in rates of litter disappearance. The short duration of the experiments, however, would make it premature to conclude that spiders are unlikely to have an effect on rates of litter decomposition. The statistical analyses of Kajak & Jakubczyk were probably too conservative. Because samples on each date were independent of each other, the appropriate model for analyzing the data is a two-way ANOVA, with time as one of the treatments. Destructive sampling of the experimental units at different times makes such an analysis possible without falling into the trap of temporal pseudoreplication. Nevertheless, the large amount of within-treatment variation in decomposition rate (Table 8.6) suggests that a more appropriate analysis would also fail to reveal statistically significant effects.

Kajak & Kaczmarek (1988) conducted a similar experiment several years later. The longer duration of this study, which lasted from May through November, was a major improvement in design. Spiders accounted for >70% of the predators sampled by pitfall trapping. Predatory arthropods were trapped >20× more frequently in the open enclosures (daily capture rates for open and closed cages were 1.56 ± 0.28 and 0.05 ± 0.04, respectively). Sampling variation again was high, but the results were similar to those of the previous experiments. Collembola densities were higher in enclosures from which spiders and other macrofaunal predators had been excluded (Fig. 8.9). The investigators report (but did not present data) that in soil cores from which predators had been excluded, litter decomposition was slower and final humus content of the soil was higher.

Caution should be exercised with respect to possible cage effects when planning similar experiments in the future. In the first year of Kajak & Jakubczyk's studies the mean density of Collembola in the open cores, which were completely exposed to predators, was similar to the density of Collembola in the closed cages, which supposedly were relatively free of predators. This similarity was most remarkable after 50 days, when partially enclosed soil cores had densities of Collembola much lower than either the completely open or completely caged cores (Table 8.6). Unfortunately, no estimate of sampling variation is given for the mean values of the open controls, even though 30 replicate cores were analyzed. Thus one can only guess at the significance of the similarity, but it is disconcerting that the open control more closely resembles the closed cage than the partial cage, which allowed predators to enter. Is the observed effect of predators amplified by the possible tendency of a small, open cage to concentrate predators long enough that they feed more intensively than usual on small patches of Collembola? This possibility is no minor problem, and weakens the generality of the

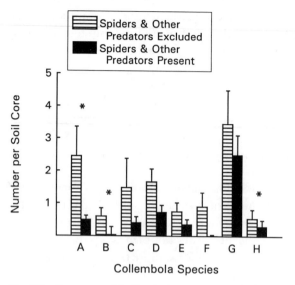

Fig. 8.9 Response of Collembola populations to exclusion of spiders and other macrofaunal predators from 15 cm × 78 cm² soil cores. Mean number ± s.e. An asterisk (*) identifies differences that are statistically significant ($p < 0.05$, simple *t*-test uncorrected for multiple comparisons). A = *Isotoma viridis*, B = *I. notabilis*, C = *Sphaeridia pumilis*, D = *Folsomia quadrioculata*, E = *Ceratophysella armata*, F = *Hypogastrula manubrialis*, G = *Onychiurus armatus*, H = Sminthuridae juveniles. (From data in Kajak & Kaczmarek 1988.)

results, particularly since the design did not include open controls in later experiments. The basic approach of Kajak and her colleagues is promising, but future research of this type should include open controls and larger experimental units.

Synopsis

This chapter goes beyond the gentle plucking of the ecological web that characterized preceding pages, and examines field experiments relevant to uncovering indirect effects. This approach requires a more persistent plucking, which demands a willingness to maintain the experimental treatment for a long time. Such long-term press perturbations yield results that sometimes resist straightforward interpretation, and that are most fruitfully understood in the context of additional information about the system. Not all of the experiments discussed in this chapter satisfy the requirements of a well-executed press experiment, but few adequate press experiments with complex communities have yet been performed. Thus it is instructive to consider the range of studies conducted to date that provide evidence on indirect effects in the ecological webs of spiders.

Intraguild predation (IGP) – predation between species of the same predatory guild – is potentially important among spiders, particularly among wandering species. In most terrestrial communities IGP between spiders and other generalist predators is also likely to influence the dynamics of many species through both direct and indirect interactions. Generalist predators can affect each other through competition and/or IGP, and can have direct and indirect effects upon prey populations. A long-term field experiment with a desert community, in which the most abundant scorpion species was removed, uncovered no evidence of exploitative competition for prey between scorpions and spiders. The investigators concluded that spiders were more abundant in the scorpion-removal plots because of reduced IGP, but the fact that experimental and control areas were not interspersed unfortunately weakens the ability to generalize from this long-term press experiment.

Two field experiments have examined interactions between lizards and web-spinning spiders on Caribbean islands. One experiment was a pulse experiment, in which lizards were removed from replicated fenced plots, and were then reintroduced into half of the areas. The manipulation produced a three-fold difference in lizard densities (the 'no-lizard' plot had some at the end of the study because of immigration and the difficulty of finding all lizards in a plot at the start). Aerial insects were more abundant in plots with more lizards, most likely because the lizards consumed large numbers of juvenile orb-weaving spiders, leading to a decrease in older spiders that foraged higher in the vegetation where the aerial insects were sampled. Insects on the forest floor were lower in the enclosures with higher lizard densities. Another study, a long-term press experiment in which lizards were removed from fenced plots for three years, uncovered no positive effects of lizards on abundances of aerial insects. However, this study did uncover clear evidence of a negative effect of lizards on orb-weaving spiders due to depletion of prey. In addition to exploitative competition for prey, predation by lizards was found to have a significant negative effect on spider numbers (details discussed in Chap. 5), and was judged to be the more important interaction.

Several studies have searched for indirect effects propagated across more than two trophic levels. Two different experiments in salt marshes have shown that fertilizing grasses can affect spider populations. In one study the density of the herbivore trophic level was boosted, although populations of a major planthopper did not change. Evidence suggests that spider populations showed a numerical response to those herbivores that did increase in numbers, which increased pressure on the planthopper population and prevented its density from increasing. Another set of experiments in a different salt marsh also produced evidence that spiders responded to increased planthopper numbers following fertilization. Spiders apparently had an impact upon planthoppers in some, but not all, years. A third long-term fertilization experiment, conducted in an agroecosystem, resulted in no change in density or species diversity of spiders, though the investigator concluded that regular fertilization led to reduced total spider biomass. Certain ambiguities in the statistical analysis detract from the certainty of the conclusions, but it is clear that fertilizing the meadow community failed to produce as large an impact upon spiders as was observed in the salt marshes.

Two field experiments are discussed that seek to determine whether the number of spiders influences the trophic level at the base of the food web. In one study different numbers of grasshoppers and a common orb weaver were added to small field enclosures. This short-term pulse experiment produced a negative correlation between the average number of spiders in each cage and the fraction of plant matter destroyed by grasshoppers. However, densities of grasshoppers were an order of magnitude higher than natural densities, making it difficult to extrapolate the results to the natural community. The other study was also conducted with a grassland ecosystem, but with the detritus-based food web. Wandering spiders are a major component of this community. The investigators indirectly manipulated densities of spiders and other predators of the ground layer by altering their accessibility to bags containing cores of soil, grass and litter. Completely closed bags had higher numbers of Collembola, but none of the treatments differed significantly in rates of litter disappearance. The investigators collected indirect evidence suggesting that a reduction in spider predation produced the increase in Collembola densities. However, conclusions on the net effect of spider predation on decomposition rates are difficult to draw. Though a press experiment in basic design, the study was probably too brief to have produced the expected impact of increased Collembola numbers on the rate of litter decomposition.

9 · *Spinning a stronger story*

Metaphors, models and paradigms

Reasoning by means of analogies is risky. The metaphors of the preceding pages arose as innocently concocted literary devices and nothing more; they were not intended to parallel nature's deeper structure. Yet stylistic creations spring from particular world views, which filter results of scientific investigations and gaze more favorably on some approaches than others. I have tried to shape the spider as a *model terrestrial predator*, who acts out the script of a drama staged and casted in the complexities of the *ecological web*. As interpreters of this theater, we cannot sit back and watch passively, because we are ignorant of what occurs behind the scenes, offstage. In order to understand the hidden scripts, we must jump onto the stage and challenge the actors – probe and poke, remove their masks and look behind the props. In short, we must rely on *field experiments*. My biases have been clear. They merit scrutiny, because they color my interpretation of what has so far been learned of spiders in their ecological webs.

Ecological dramas

The spider persona
What have we discovered about the spider persona? The spider is usually hungry. When given more prey than it normally encounters, the spider grows faster and produces more eggs. Sometimes it will wander less when food becomes more abundant, but at other times it apparently behaves independently of its recent feeding history. Despite the fact that food is frequently a limited resource, spiders do not commonly compete with other spider species for prey. Spiders avoid competition not because they have assumed different roles, but because their densities are usually below levels that lead to significant intraspecific exploitative competition. Thus niche partitioning is not the best explanation for why

interspecific competition among spiders is rare. In one intensively studied case competition clearly was present, but it is unlikely that observed niche differences between the two competing species are co-evolved adaptations to past competition.

Why spider densities are often below competitive levels is a straightforward question with no simple answer. One territorial species spaces itself in a pattern that minimizes exploitative competition for prey. Many abiotic and biotic mortality factors have been identified for spiders. Their combined effects must contribute to the absence of strong competition in many spider populations, but so far no experimental study has been conducted that would demonstrate whether one natural enemy or abiotic source of mortality is decisive in preventing resource competition in a particular spider community. The ecological webs of most spiders are likely too intricate for any such simple pattern to emerge.

Intriguing paradoxes define the spider persona as presently understood. The spider is food limited, yet does not compete for prey. Thus spiders do not consume enough prey to lower the prey availability substantially for other spiders. This suggests that spiders do not limit population densities of their prey. Furthermore, the foraging behavior of spiders and their life history characteristics lead to the prediction that spiders should not regulate prey populations, i.e. should not inflict pronounced density-dependent mortality on their prey. Thus individual species of spiders should not be good biocontrol agents in agroecosystems. Indirect evidence, based on estimates of energy flow through the entire community, and direct evidence from field experiments contradict these expectations. However, experimental studies are too few in number to support firm generalizations about the role of spiders in limiting insect numbers. Results so far lead to an enigmatic portrait of the spider persona; its role is clear in some scenes, but only dimly perceived in other acts. Many scripts it follows are still obscure.

The ecological play
Perhaps the metaphor of the ecological play is not useful. A textbook for beginning students cautions them that 'the lives of organisms are played out for their own benefit, and not for the purpose of fulfilling some role in the ecosystem as an actor fills a role in a drama' (Ricklefs 1983). Ricklef's point is valid. It is the actors themselves that develop the scripts as they play out the evolutionary drama in the ecological theater (Hutchinson 1965), but they are constrained by the available props,

supporting actors and past performances. Whatever the origin of the script, though, organisms behave as if they were following one. Their performances may not be identical from production to production, but there is a predictable pattern to the way they play their roles. Whether or not a director exists somewhere way offstage is a question best left to metaphysics.

Numerous renditions of the same play are produced throughout the earth's ecosystems. A basic problem with the metaphor of ecological theater is uncertainty over whether or not these versions are variations of the same underlying drama. In order to determine whether a unifying theme exists, we must repeat studies on similar but different systems – in different years and different places. This requirement by itself poses a formidable challenge, but we also face another: to define the personae in the ecological drama we have chosen to study. Who are they, how many actors are we to follow, which ones should we ignore? The difficulty with the metaphor of the ecological play is not whether organisms follow a script, but rather, how do we recognize what performance we are attending and whether similar performances of the same drama are produced elsewhere? The problem is to identify an appropriate *ecological model*.

Model species and model systems

Ecologists frequently conceive of models as mathematical constructs designed to embody basic characteristics of individuals, populations or ecosystems. Such models are hoped to *represent* the essential properties of real systems. Models need not be mathematical constructs. Laboratory populations, for example, can be viewed as models of real populations. Turnbull (1964) demonstrated that the frequency with which the common theridiid, *Achaearanea tepidariorum*, re-located its web within a laboratory room depended upon its feeding history at a web site. Whether or not this behavior is *representative* of wild populations can only be answered with field experiments. The question is somewhat moot for this example, because the focus of Turnbull's study, the common house spider, was living in one of its natural niches. Some aspects of behavior can be studied in the laboratory and extrapolated to outside conditions, but field experiments in ecology most frequently address the question of whether a particular interaction occurs within the context of the natural community, i.e. within the constraints set by all other abiotic and biotic factors of the natural ecosystem. It is unlikely that one will successfully create a laboratory microcosm containing spider populations, their prey and natural enemies that would be representative of real populations. Rather than struggle to create a representative community in the laboratory, it is usually more productive to attempt the field experiment.

Results of field experiments will be most useful in building theory if they are conducted with organisms that represent a wider class of species. Is the spider a good model terrestrial predator, as I proposed elsewhere (Wise 1984a)? Grabbing an individual at random from a collection of terrestrial predators (excluding the mesofauna of the soil) would often yield a spider. But is it then safe to assume that the spider's foraging behavior, population dynamics and interactions with other species are representative of patterns exhibited by all other terrestrial predators – vertebrate and invertebrate? This is unlikely to be true, but it remains valid that understanding spiders in their ecological webs would make a major contribution towards understanding how predators function in terrestrial systems, particularly if generalizations based upon spiders are restricted to predators that occupy intermediate positions in the food web (Schoener 1989). As proposed previously (Wise 1984a), the spider is *a* good model terrestrial predator; it would be foolhardy to search for *the* model predator.

A good model must be characterized fully. Does *the spider* constitute a single persona in the ecological drama? The pattern emerging from the last eight chapters suggests not. Although a paucity of data exists for many spider families, enough has been collected to show that many spiders differ sufficiently in behavior and circumstance that they should be assigned separate leading roles. The clearest argument for caution in depicting a single characteristic ecological role for spiders comes from one of the central conclusions of this book: the failure of the competitionist paradigm to explain patterns in spider communities. Evidence of current interspecific competition has not been found frequently when it has been searched for with field experiments, but practically all of the data come from investigations of large, conspicuous web builders. We have no good experimental evidence on whether or not species of wandering spiders compete for prey. Indirect evidence hints that exploitative competition could be more frequent in these groups. For example, many of the spiders implicated in biocontrol of agricultural pests are not web builders. Wandering spiders of the litter layer of forests and grasslands live in close physical association with many prey populations. These spiders may have a large impact upon their prey populations and consequently may suffer from exploitative competition. Not until we have resolved this issue and others equally important will our understanding of the spider persona be reasonably complete. It is already apparent that the notion of *the* spider persona should be abandoned in favor of individual portraits of more carefully defined groupings, likely to be drawn along family lines. Lycosids and linyphiids undoubtedly play quite different roles. Among the cursorial species, night-active clubionids and anyphaenids may play different parts to oxyopids and salticids, which depend upon well-developed eyesight to maneuver. Or these groups may play roles that are remarkably similar. We will not know without further research.

Attempting to characterize model species may prove limiting, particularly if the ultimate goal is to characterize patterns at the community level of organization. The tradition of defining communities horizontally, i.e. within a trophic level, and restricting membership along taxonomic lines, is central to the controversy of interspecific competition, and is an approach necessarily employed in part of this book. McIntosh (1987) warns that the dilemma of deciding among model systems defined along taxonomic lines 'recalls the widespread concern about the 'new' community ecology of the 1960s and 1970s – that the avian tail was wagging the ecological dog.' This point is well taken, but the danger appears small that models spun out of arachnological arrogance will ensnare unwary ecologists. Continued vigilance would be prudent, though. Model organisms are forever seeking an audience. For example: advertising for a monograph on lizard ecology boldly asserts that 'research on lizards is helping to establish a new and more versatile paradigm for ecology . . . lizards may well be challenging birds as the model organism in ecology' [advertisement for Huey, Pianka & Schoener (1983) on back cover of *The American Naturalist* 121(5)]. Community ecology is leaving behind earlier attempts to find single-cause explanations for patterns in communities defined along taxonomic lines; emphasis is being placed increasingly on comprehending interactions both within and between trophic levels. Thus the problem of defining a model species, or groups of species, shifts to one of finding communities to study that are representative of a wide class of systems. One criterion used in selecting such model systems should be amenability to field experimentation.

Paradigm of the field experiment

Arguments will continue about the qualities of appropriate model systems for ecological research, but any final resolution will depend upon the ability to generalize, which ultimately depends upon the type of data that can be collected. Debates over the appropriate scope of hypotheses, and how to collect data in order to test them, are ultimately more important than quibbling over which types of species collections constitute the more powerful model systems. The issues are linked, though, because some model systems are more amenable than others to field experiments. This book has focused relentlessly on field experiments for two reasons: (1) controlled, replicated perturbations of natural conditions will ultimately reveal more than detailed, non-manipulative studies for those systems amenable to field experiments; and (2) many spider populations have proven to be good subjects for the experimental paradigm.

As ecologists focus more on global questions, the relevance of replicated, controlled field experiments will be increasingly questioned. Understanding processes over scales larger than the boundaries of a field experiment requires other approaches. Not all ecological interactions can or should be studied by controlled field experiments. In many systems manipulations cannot be accomplished cleanly, the spatial and temporal scales of interest do not mesh with the experimental paradigm, or it would be unethical to perturb the system. A pluralistic approach to formulating and testing ecological hypotheses promises to be the most productive (e.g. Diamond 1986, McIntosh 1987). A multi-faceted approach is most productive even for systems that are well suited to field experiments, but the distinction always should be maintained between discovered patterns that *generate hypotheses* and planned field experiments that constitute *tests of hypotheses*.

Experiments with spiders have taught us much about the usefulness and limitations of field experimentation. It will be instructive to examine some of these lessons before closing with suggestions on what must yet be done to spin a stronger story about the spider's ecological web.

Plucking the web

Spiders as tools of the experimentalist

Spiders offer many advantages to the experimentalist. Their intermediate size and often high abundance make it feasible to establish conveniently sized study sites that can be perturbed and sampled with less logistical effort than would be required for experiments with a larger predator. Other invertebrate groups offer the same advantages of size and abundance, but no other terrestrial predator has the attractive features of the web-spinning spider.

The web and the semi-sessile behavior of its owner make the spider an almost unique tool for the terrestrial animal ecologist. Because many spiders do not move web sites frequently, it has been feasible to establish populations of different densities and species mixtures without constructing barriers to movement. The fence is probably the most serious source of artifacts in field experiments, and its absence is a major asset. Once in place, the web spider presents other desirable features: (1) the experimenter can supplement the prey of individuals by adding insects to the web; (2) feeding behavior can be observed directly, and in some species past feeding can be inferred from prey remains left in the web; (3) behavioral interactions with conspecifics, other spiders and natural enemies are readily observable in the arena that the web provides; and (4) fecundity in many species can be measured directly by plucking the egg case from its owner's web.

Spiders are not the experimentalist's panacea. Like all arthropods, spiders molt, making it extremely difficult to follow marked individuals other than adults. Manipulating unmarked individuals in a system open to immigration can complicate interpretation of results, particularly if the investigator has established treatments of different population densities; a good example is found in my research on the filmy dome spider (Wise 1975; Chap. 5). Although immigration rates of this linyphiid were low (one immigrant/unit/month), they still led to ambiguity because the experiments were designed to detect changes in experimentally established low-density populations. Higher migration rates can cause logistical problems in removal experiments; immigration rates of web-spinning spiders into open plots can be high enough that frequent removals are needed to maintain 'zero'-density treatments (e.g. Horton & Wise 1983, Riechert & Cady 1983, Spiller 1984b). Of course populations of webless spiders cannot easily be manipulated without using fences, although it should be feasible to manipulate some cursorial species without erecting barriers. For example, some thomisids are closely associated with flower heads, and could perhaps be manipulated successfully in short-term experiments. Although wolf spiders must be confined with barriers, ecologists who manipulate lycosids are at least grateful that the female carries the egg sac and recently hatched spiderlings with her.

Insights derived from experimenting with spiders

Field experiments with spiders have modified some preconceptions about the roles these arachnids play. Results of experiments have largely discredited explanations of spider community structure based upon the competitionist paradigm. Perturbations of spider densities have shown that web spiders, though frequently food limited, do not experience exploitative competition for prey. This latter finding is not controversial nor perhaps surprising in retrospect. However, before the experiments had been conducted, most ecologists probably would not have expected a shortage of prey to act as a density-independent limiting factor.

Conventional wisdom suggests that individual spider species should not be good biocontrol agents, a prediction superficially at odds with the widespread belief that spiders, because of their abundance, should control insect populations. A recent field experiment demonstrated that the entire spider community can have a substantial impact upon pests in an agroecosystem (Riechert & Bishop 1990; Chap. 6). This finding can be reconciled with both types of conventional wisdom by recognizing that individual predators that fail to display strong density-dependent responses to prey populations can together form a predator assemblage that exerts substantial density-independent mortality upon prey populations. Somewhat surprising, though, are field experiments which suggest that a single species of clubionid may be instrumental in controlling certain pests (Mansour *et al.* 1980b, Mansour & Whitcomb 1986, Mansour 1987; Chap. 6). This is a counter-intuitive conclusion that merits more experimental scrutiny.

Counter-intuitive results are always exciting, because they force us to scrutinize cherished theories and abandon habitual ways of interpreting patterns in nature. The major justification for performing field perturbations, however, is not to uncover patterns that run counter to intuition, but rather is to test hypotheses in as direct a manner as possible. Of course, it is not only results of field experiments that challenge dogma or support hypotheses. Many approaches exist in ecology, three of which traditionally are classified as being experimental: laboratory experiments, field experiments and natural experiments. Diamond (1986) reviews their strengths and weaknesses as tests of hypotheses in ecology. As emphasized earlier, laboratory studies can elucidate phenomena, particularly aspects of physiology and behavior, but their relevance to understanding community-level processes is that of a *model*, not of an experiment in the sense of a field experiment. Diamond makes many excellent points about the strengths and limitations of the three approaches, but **only the field experiment deserves the appellation** *experiment* **in population and community ecology.**

The natural experiment

The natural experiment relies upon 'natural perturbations occurring in the field' (Diamond 1986). The natural experiment does not rely upon the designed manipulation of one or more factors among a set of habitat units, in which replicate units have been assigned treatments at random. Consequently, in the natural experiment no way exists to estimate the probability that the observed pattern is caused by variation in unknown factors. Such correlations do no merit the title of *experiment*. This criticism is not a regression into semantic quibbling, a seductively attractive displacement activity that occasionally afflicts ecologists. The issue is not 'mere semantics': to term a 'natural experiment' an *experiment* is misleading because the name suggests a power to test hypotheses that this type of non-manipulative correlation does not possess. A natural experiment can suggest a hypothesis, but by itself can never constitute a test of that hypothesis (Wise 1984a, Hairston 1989).

An excellent example of this potentially creative connection between a natural experiment and a field experiment comes from research on lizards and spiders in Caribbean islands. Schoener & Toft (1983a) found a pattern across islands suggesting that lizards have a negative impact upon the abundance of web-spinning spiders. The long-term field experiment of Spiller and Schoener (1990a) confirmed a causal connection, and also provided information on interactive mechanisms. The results of the field experiment do not prove that lizards were the direct cause of the pattern observed in the natural experiment, because other factors could have been more important on the particular islands surveyed. A more direct confirmation would have come from manipulating lizard densities on the islands that were identified as the natural experiment. Spiller & Schoener have not removed lizards from islands with spiders, but Schoener has followed the persistence of spiders introduced on to islands with, and without, lizards (Schoener 1986; Chap. 5).

Successful demonstration of a lizard-effect in the experiment of Spiller & Schoener does not justify accepting the natural experiment as a means of testing hypotheses. Coincidence between a correlative pattern and the results of a controlled field perturbation does not alter the nature of the natural experiment; it still constitutes indirect evidence. However, combining the two approaches strengthens generalizations about interactions between lizards and spiders on Bahamian islands.

What if the field experiment had not uncovered evidence that lizards affect spider numbers? Explaining the correlation of the natural experiment would have been made more difficult. If a field experiment fails to confirm the pattern of a natural experiment, one could elect to pursue more detailed studies, which might suggest alternative explanations that could then be put to experimental tests, or one might decide instead to conduct additional field experiments, which might eventually uncover results consistent with the natural experiment.

What if the natural experiment has not been done? If nature has not performed the experiment, i.e. not provided us with the correlation, then the manipulative field experiment is the only available technique. It must be fairly common that variation in many factors between different sites obscures simple correlative patterns. Evidence of food limitation in spider populations provides a good example. Indirect evidence leads to the prediction that spiders are food limited. However, within-habitat variation in food supply may be too small for strong correlations between prey abundance and spider population parameters to emerge. Direct augmentation of prey abundance in a field experiment would then provide the most direct, convincing evidence of food limitation, and can also yield an estimate of its magnitude. Whenever possible, of course, using field experiments to confirm the interpretation of an observed correlation is the most productive approach [cf. Martyniuk's (1983) study of prey availability at web sites and residence times; Chap. 2].

Diamond (1986) points out that natural experiments can encompass temporal and spatial scales unattainable with manipulative field studies. Such broad correlations are an important source of insight; field experiments can never substitute. Nevertheless, terming such correlations *experiments* obscures the strengths and weakness of two quite different approaches. Distinguishing between them would remind both seasoned and nascent ecologists of their discipline's pluralism, and of the need to separate the erection of ecological hypotheses from the testing of their validity.

Recognizing a clearer distinction between the two approaches would also force us to define more clearly the *scales* over which hypotheses in ecology are formulated and tested. One important aspect of the problem of scale is that of variability: how temporally and spatially *unique* are the processes that ecologists deem worthy of study? This question confronts the empiricist as issues of *model systems*, *experimental design* and *negative results*. The first has just been discussed; the other two, which are equally important, are treated next.

Experimental design

Hurlbert's (1984) critique of field experiments in ecology has forced ecologists to focus on several critical aspects of designing and interpreting manipulative field studies. The most significant contribution of his paper is its warning to avoid the trap of using variation between sub-samples, within either single or pooled replicates, to estimate the error variance for hypothesis testing, a mistake Hurlbert termed *pseudoreplication*.

Pseudoreplication

Pseudoreplication has plagued all types of field experiments, including those with spiders. The most troublesome version is *simple* pseudoreplication, in which only one plot per treatment exists. *Sacrificial* pseudoreplication occurs when an investigator has established a proper replicated design, but pools sub-samples from all replicates for the data analysis. Redoing the statistical analysis is an easy remedy for the latter type of pseudoreplication. The other error can be corrected only by re-doing the experiment. It is unfortunate, but in a simply pseudoreplicated design one can use statistics to compare plots, but not to test for significant differences *between treatments*, i.e. not to *test* a hypothesis.

I have called attention to pseudoreplication in my experiments and those of others because the problem reflects more than statistical *naïveté*, and deserves some focused scrutiny. Though we cannot escape the truth that ignorance of the basics of experimental design and principles of ANOVA is the ultimate explanation for the phenomenon, most of us reformed pseudoreplicators would like to attribute our past mistakes to causes more profound than simple stupidity. One cause of pseudoreplication is the need to establish experimental plots that are as large as possible in order to maximize the realism of the experiment. Logistical constraints then limit the number of replicate plots that can be handled in a single experiment.

Use of large plots improves chances that representative amounts of spatial heterogeneity will be included and edge effects will be reduced. Ecologists ensure that plots resemble each other as much as possible by locating them close together and by selecting areas that are similar in vegetation, exposure, etc. Nevertheless, any two such plots will be different, no matter how carefully chosen; this variation is the source of the experimental error against which the investigator compares differences attributed to the experimental treatment. Experimental error, the inherent background variation in the system, cannot be estimated without replicate plots. Knowledge of initial conditions would provide one indication of how similar the plots are, but by itself cannot substitute for spatial replication. If undetected factors affect the plots differently after the experiment has started, the investigator may unwittingly conclude that a difference appearing between plots reflects a treatment effect when in fact the imposed treatment had no impact. Replication of plots and physical interspersion of treatments are the only remedies against what Hurlbert (1984) terms 'nondemonic intrusion' of chance events during a field experiment.

Even with replication of control and treatment units, detecting treatment effects may be difficult if spatial heterogeneity across the study area produces a substantial error term. Familiarity with initial plot conditions can reduce the impact of this variation. Recording values of the response variables in control and treatment plots before perturbing the system makes it possible to analyze effects of the treatment on an index of change. Error variance for this index should be less than that calculated for values of the parameters known solely at the end of the experiment. Hairston (1989) argues strongly for knowing initial conditions: '. . . to carry out a satisfactory experiment, one must have full knowledge of the conditions existing before the experiment is begun.' Hairston overstates the case somewhat, because one can legitimately analyze final conditions as long as treatments have been assigned to experimental units at random. However, limiting the analysis to endpoints of the experiment and retreating to a defense based upon statistical purity could be dangerous. It is always possible that an initial bias could exist by chance, particularly if the number of replicates is small. Low replication is usually true of ecological field experiments. Thus ecologists would do well to follow Hairston's advice and collect as much baseline data as possible before beginning the manipulation.

Additional information about the plots aids in interpreting results and planning the experiment, particularly if only a few replicates are feasible. A good example comes from the competition study with the two *Argiope* species (Horton & Wise 1983). The design had two replicates (blocks) per treatment, and in the first year's experiment a plot in one of the blocks had particularly tall vegetation. One apparent effect of spider density was to increase the mean height of the webs in the vegetation.

Unfortunately, the plot with high vegetation was assigned one of the high-density treatments. We attempted to remove the possible effect of this unusual plot when we repeated the experiment the following year. Treatments were still assigned at random, but with the proviso that the plot that had particularly high vegetation the first year not be a high-density treatment. Spiller (1984a,b) minimized the effects of vegetation in his ANOVAs by expressing mean web height in a plot as a fraction of the vegetation height.

Knowledge of initial conditions and gathering of basic natural history data are particularly critical when replicates are few. This condition is likely to prevail in ecological field experiments, even now that ecologists are more conscious of pseudoreplication. Replicates will usually be few in number, because establishing additional replicate plots in order to enhance error degrees of freedom comes at the expense of making plots so small that the processes studied within them become artifacts of the experimental design. It does no good to renounce the sin of pseudoreplication if one unwittingly turns to embrace the folly of pseudorealism.

Pseudorealism

Ecologists who perform field experiments imagine themselves to be the purest of empiricists because they manipulate the actual system – not an artificial laboratory model – and they do not rely on the *a posteriori* arguments that the sifted evidence of the natural experiment makes inevitable. Such purity is only a relative virtue. Two major impurities of the experimental approach are the need to select a small portion of the ecosystem for experimentation, and the frequent requirement to erect barriers to migration. Issues surrounding these two aspects of field experimentation are interrelated.

Barriers to movement across plot boundaries

Barriers to movement introduce an element of artificiality. Cages are the most extreme; thus Rypstra's (1983) study of the effect of food supply and substrate upon interactions between web-building species is really a laboratory-type experiment conducted in the field (Rypstra made no claim that her study mimicked a natural population). Other investigators have also conducted behavioral studies of spiders within field cages (e.g. Riechert 1981, Olive 1982). These studies were more conveniently performed outdoors, where abiotic conditions are more natural than in the laboratory. No simple way exists to establish a control for cage effects in studies such as these, because allowing spiders to emigrate would make it difficult to study the behavioral interactions the experiment was designed to examine.

The most common cage experiment in community ecology attempts to duplicate natural conditions more closely. For example, Hurd & Eisenberg (1990) used cages to study population interactions between naturally occurring prey populations and cursorial predators (a lycosid and a mantid) that had been added to the cages. The pasture community trapped inside was intentionally left as natural as possible because the study's goal was to make inferences about processes occurring outside the barrier. Nevertheless, problems of extrapolating to the wider system are present, particularly because the cages were small (1 m²). However, the short-term nature of the experiment (10 days) lessened the probability that cage effects would be severe. Given the density and probable home range size of the wolf spider, cages at least an order of magnitude larger in area would be required for a longer-term experiment.

Cages have also been used to exclude predators of spiders that cannot be manipulated directly. In most field experiments with spiders, though, cages have not been used as frequently as partial barriers to migration. Most enclosures have been made of low fencing, which permits movement of flying insects across the plot boundary.

Enclosures have not always been used when manipulating mobile arthropods. Barriers might not be needed if the plots are large enough to decrease the impact of migration. Polis & McCormick (1986; Chap. 8) established two very large open plots [30 000 m² (removal) and 6000 m² (control)] for a long-term study of the impact of scorpions on spider populations. The investigators probably decided to establish such a large scorpion-removal plot in order to minimize the consequences of scorpion immigration, but unfortunately the size of the plot made it unfeasible to replicate. It would have been better (but much more work initially) to have divided an area equal to the removal and control plots into physically interspersed fenced plots of smaller size. Plots could have been made large enough to minimize effects of fencing on interactions within the system, but small enough to make replication feasible.

But how small can plots shrink before edge effects seriously cloud the results? Minimum adequate plot size is a function of the system under investigation, the scope of the question, and the duration and nature of the perturbation. Fenced plots as small as 1 m² appear to be adequate for studying competition among young *Schizocosa ocreata* (Lycosidae; normal density of 80/m²; Wise & Wagner 1992; Chap. 5), yet plots 400 × larger contained only 6–11 adult burrowing wolf spiders, which made them too small for a long-term study of predation and resource limitation (Conley 1985; Chap. 2); only one spider remained per plot by the end of the experiment. The enclosed soil cores used by Kajak & Jakubczyk (1975, 1976, 1977; Chap. 8) were very small (0.01 m²). Clearly the investigators did not attempt to include an intact segment of the community, but instead intended only to alter accessibility of Collembola and other soil/litter organisms to highly mobile predators. For a short-term pulse manipulation this restriction may cause no serious problems, but much larger plots would be required for a long-term press manipulation. The full array of indirect effects will not occur if excessively small enclosures prevent realistic population-level responses to the experimental perturbation.

Because the magnitude of fence effects ultimately depends upon the size of the experimental unit and the idiosyncracies of the system, the only remedy is to establish an open control. Absence of such a control limits the ability to generalize results of a field experiment to the natural system. For example, evidence of exploitative competition for prey among young *Schizocosa ocreata* must be considered preliminary because only fenced plots were used (Wise & Wagner 1992; Chap. 5). Evidence of exploitative competition within the confines of the plots is clear, and the experimental densities bracketed the estimated density on the open forest floor. Nevertheless, the enclosure could have influenced the manner in which the spiders interacted with their prey. Pacala & Roughgarden's (1984) experiment uncovered dramatic effects of lizard predation on spider numbers, but the design lacked an open control. Plots were so large (144 m²) that it is hard to imagine that fencing could have influenced the results, yet Spiller & Schoener (1990a) detected statistically significant effects of fencing on prey abundance in the large plots (84 m²) employed in their long-term experiment. They were able to detect fence effects because they had established open control plots.

Statistically significant barrier effects need not necessarily invalidate the experiment. Spiller & Schoener (1990) conclude that although fencing may have intensified exploitative competition for food, the extent to which insect abundance was lower in the fenced plots was within the normal range of annual variation in insect numbers. They conclude that the intensity of competition uncovered is thus not atypical.

In contrast, prey densities in the open control in the first set of experiments of Kajak & Jakubczyk (1975, 1976) suggest that cage effects may have substantially affected the outcome.

Plot size

The problem of what size plot to use centers on a few basic questions: Will enough organisms be included to mitigate influences of random effects? Is the ratio of perimeter to area large enough to minimize effects of migration upon important interactions? Is the spatial heterogeneity characteristic of the system well represented? Decisions on plot size are often inseparable from concerns about fence effects, a topic that has been addressed already. The safest, least controversial answer to the problem of what size plot to use is easy: use plots as large as feasible. But feasibility does not guarantee realism. The most direct way to examine the impact of plot size on the interactions being studied is to use a series of plot sizes. This is rarely done. Experimentalists already are struggling to obtain a reasonable number of replicates, and introducing yet another treatment is not attractive.

Closely related to the decision of what size plot to select is that of how far apart to locate them. This latter decision bears directly on the degree of generality one wants from the experiment (Diamond 1986, Hairston 1989). Field experiments are usually designed to detect a phenomenon and measure its intensity. These aims are best met if experimental error is minimized, which is accomplished by selecting plots that are near each other or that are grouped in a randomized block design (Hurlbert 1984). These features of experimental design, however, are at odds with a competing goal of field experimentation: to detect the intensity of a phenomenon within the natural variability inherent in the system. To meet this aim the investigator should select plots that are as different as possible, in order to strengthen the generality of the results. No simple formula exists for solving this dilemma; in fact, one should not expect an easy solution. The problem is part of the larger issue of how we, as a community of scientists, erect generalizations from the results of specific studies. As our generalizations and theories become more sophisticated, empiricists will become more conscious of these opposing needs when they select the size and spacing of experimental plots.

Other impurities

Ecologists studying web-building spiders have used artificial web substrates in field experiments. One goal has been to test for web-site limitation by adding web sites to the environment. In other experiments the substrate was large enough to accommodate many spiders, and was used to establish open, experimental populations of different compositions in order to examine exploitative competition for prey, interference competition and interactions with natural enemies. In these studies web-site limitation in the habitat was not being tested, and the investigators assumed that the inert material used for the web substrate did not affect prey densities nor influence the behavior of natural enemies. However, no controls were established to test these assumptions. One could measure prey abundances with sticky traps to test whether the substrates affect insect activity, but establishing controls for all other possible interactions would be difficult. Artificial web substrates are usually employed in order to establish replicated populations that can be censused and manipulated in a habitat where vegetation structure makes such manipulations difficult. Hence establishing controls for all possible effects of the substrate is unfeasible, particularly given the logistical constraints accompanying most field experimentation.

Impurities introduced by using artificial web substrates, fences and cages insure that field and laboratory experiments are not entirely distinct. Both are models of the undisturbed system. Diamond (1986) proposes that the two types of experiments be viewed as forming a continuum. Strictly this is so, but in community ecology a well-designed field experiment, even one using modified substrates or barriers to migration, is so much closer to the natural world that the corresponding laboratory experiment is most realistically placed in a category all its own. Hairston (1989) discusses field experiments in which the use of barriers leads to 'managed environments' as 'seminatural experiments,' which 'provide one possible transition between variable and otherwise difficult natural situations and the unrealistically simple environments of the laboratory.' This distinction between the 'open environment' and the 'managed environment' field experiment is useful, particularly for studies of the latter type in which there are no controls for barrier effects.

Problems associated with modified substrates or barriers do not plague experiments with spider populations on open patches or units of natural vegetation. However, even in these situations unsuspected factors can complicate interpretations, e.g. the differential impact of immigrants on populations of unmarked spiders at different densities (Wise 1975). One's interpretations will be more robust the more one knows about the fates of individual organisms and the levels of non-manipulated variables. It is always risky to isolate a piece of the natural world by imposing artificial boundaries, either by erecting barriers or simply by recognizing arbitrary open units for study. Unrecognized modifications of interactions or altering of results by migration may lead to erroneous generalizations. Nevertheless, risks of making false conclusions are potentially more severe when interpreting static or dynamic patterns in ecological webs in which components have not been intentionally perturbed in a replicated, controlled manner.

The power of negative results

A field experiment, like any experiment, can fail if the intended perturbation is not satisfactorily imposed. Thus failure to reject the null hypothesis, i.e. failure to uncover a statistically significant difference between control and treatment groups, could result from such a failure in experimental design. Such an outcome is best classified as a failed experiment, although much might be learned about the system in discovering why the perturbation, most often a reduction or augmentation of density, could not be achieved.

Failure to find statistically significant differences is frequently condemned because the experiment is judged to have yielded 'negative results.' This description of the outcome is unjustified if the experiment has been properly designed and executed, i.e. if the system has been perturbed to the intended degree, with proper controls and replication. With such an experiment, failure to reject a statistical null hypothesis is a convincing failure to confirm the underlying scientific hypothesis. 'Negative results could have more impact than those that confirm a generally held hypothesis, since frequent failure to find evidence for a phenomenon should force reformulation of theory. If a theory is worth testing, any results of adequately controlled experiments are relevant. Why test the theory if only positive results are interesting? A natural and valid disappointment over negative results may accompany the failure to produce an example that confirms theory. Any disappointment, however, should be with the theory, not with the experiment' (Wise 1984a). Uneasiness over negative results may have contributed to the failure of ecologists to recognize pseudoreplication in their statistical analyses. After having finished a laborious study, one may be unwittingly predisposed to overlook inflated degrees of freedom if the scientific community pays little heed to non-significant F-values.

Ecologists may be uncomfortable with negative statistical results because one can never confirm the null hypothesis. It can never be affirmed with certainty that treatment and control plots are samples from the same underlying distribution. Because treatment effects could have gone undetected due to insufficient replication, the possibility always exists that one will falsely accept a statistical null hypothesis, i.e. commit a Type II error. This risk is basically no different, though, than the gamble taken when we reject the statistical null hypothesis and accept the scientific hypothesis that predicts the treatment effects. We could be falsely rejecting the null hypothesis, committing the Type I error with which ecologists are more comfortable. Ecologists readily calculate the α or significance level of their statistics, but rarely determine β, the probability of committing a Type II error. They neglect β because it cannot be calculated without specifying a specific alternative hypothesis (Sokal & Rohlf 1987). Most ecologists do not entertain quantitatively defined alternative hypotheses, i.e. hypotheses that state the amount by which treatments and control differ. Instead, the alternative is simply that the treatment mean does not equal the control mean, or that it differs in a particular direction.

Toft & Shea (1983) convincingly argue that ecologists should be aware of the statistical power of their tests. The power, which equals $1 - \beta$, is the probability that the test will correctly reject the null hypothesis. In order to calculate this quantity, investigators must decide upon the magnitudes of the effects they hope to detect (Toft & Shea 1983). Rotenberry & Wiens (1985) amplified on these points and suggested that researchers calculate Cohen's (1977) comparative detectable effect size (CDES), which is defined for a certain sample size, with $\beta = \alpha$. If one fails to reject the null hypothesis, one can then state that at the β confidence level, the difference between treatment and control groups is no larger than CDES. It is then up to the investigator and other ecologists to decide the magnitude of a meaningful CDES.

A disturbing pattern emerges from my review of field experiments with spiders: experimenters (including me) that did not uncover significant treatment effects failed to consider the possible power of their statistical tests, nor did we calculate CDES values or something equivalent. A recent comprehensive review of field experiments in ecology (Hairston 1989) makes no mention of power analysis, undoubtedly reflecting the fact that none, or very few, of the reviewed papers considered it. Examination of other reviews of field experiments would likely reveal the same omission.

If we are to utilize field experiments most effectively as tests of hypotheses, we must more consciously focus on the magnitude of the effects uncovered when differences are statistically significant, and the CDES of a test when the null hypothesis is accepted. This approach will foster development of quantitative ecological theory. More concern with the power of statistical tests will also focus attention on the issue of adequate sample sizes in field experiments. Eberhardt & Thomas (1991) argue that 'the unquestionably important issue of "pseudoreplication" is overshadowed by the very small sample sizes used in ecological field experimentation.' Increasing recognition of the problems of replication in field experimentation should be accompanied by a greater appreciation of the information content of studies that yield statistically insignificant effects. Such results should no longer be labelled 'negative'; they will, in fact, make increasingly positive contributions to testing and developing quantitative hypotheses if attention is given to the power of the tests being employed.

Poking the web

Field experiments serve not only to test hypotheses, but can also yield unexpected results that expand basic knowledge of the system. Uncovering indirect effects is a prime example, e.g. the decrease in insect numbers caused by the release of spider populations from lizard predation (Pacala & Roughgarden 1984; Chap. 8), or the consequences for spiders and planthoppers of fertilizing salt-marsh grasses (Vince et al. 1981, Döbel 1987; Chap. 8). This potential payoff of field experimentation is perhaps obvious, but the extent to which experimentation should be uncoupled from hypothesis testing is controversial. Although rarely done and practically never admitted in research publications if it has been done, perturbing the system with no specific hypothesis in mind could yield worthwhile results. Diamond (1986) points out that '. . . field experiments of the 'poke-it' type on even the best studied communities may still yield major surprises . . . obsession with hypothesis testing should not blind us to the possible value as well of some manipulations unconstrained by prior hypotheses.'

Considering the expenditure of time and effort required to conduct most field experiments, it might seem foolish to perturb a system without first entertaining a well-reasoned hypothesis. Most ecologists would give such cautious advice. Is it not more rewarding to use an experiment to test a specific hypothesis? – unexpected results may turn up anyway. For example, almost a fifth of the competition experiments reviewed by Connell (1983) uncovered evidence of facilitation (Vandermeer, Hazlett & Rathcke 1985). Strict adherence to hypothesis testing is not as widespread as might be believed. Because not enough is known about the system beforehand in order to support a *precise* hypothesis, many experiments are of the poke-it type even though stated in terms of a hypothesis test. This tendency is likely what prompted Hairston (1989) to argue that a necessary element of experimental design is thorough knowledge of the initial conditions of the plots before the experiment is started, because '. . . a prediction requires a clear statement of the conditions that must exist for it to be confirmed . . .' and '. . . a satisfactory hypothesis must, at the very least, contain one or more overt or implied predictions.' Hairston adds that a '. . . curious ecologist would not perform the simplest manipulation if there were no expectation of some result, however nebulous . . .' The fuzziness of the expectation is what separates the hypothesis test from the poke.

Most ecologists who utilize field experiments in their research probably share Hairston's view. We also would agree on the importance of gathering as much natural history information as possible, in order to interpret thoroughly and accurately the results of the experiment. Natural history data usually are collected without a hypothesis in mind. Is it not reasonable, then, to view the field manipulation as an additional approach to obtaining basic information about a system? For example: we may want to know whether an abundant spider limits population densities of its prey. Detailed data could be collected on feeding preferences of the spider, relative densities of the spider and its prey populations, other natural enemies that feed upon the spider's prey, etc. We could then erect a hypothesis about this particular situation, i.e. we could make a prediction based upon the accrued natural history information. Of course we could make several different predictions, based upon what importance we attribute to responses such as increased resource competition among the prey populations when released from predation pressure, increased mortality from other predators in the absence of the spiders, inability of the spider population to show strong functional and numerical responses because of territorial behavior, etc. In this situation it might be easier and more straightforward to perform a long-term removal experiment. Insights could result that the gathering of detailed non-experimental data would probably never furnish.

I am not arguing for field experimentation in the absence of preliminary information on the system, nor am I advocating that we abandon the collection of supplementary data during the course of the experiment. Such data aid in interpreting results of the manipulation, particularly when one expects indirect effects to surface. Pluralistic approaches to ecological questions will always yield the most information. I am advocating wider recognition of the insights to be gained from 'poke-it' type experiments. Recognizing which systems will reveal the most in response to such an approach will require careful thought, but there should be wider recognition that it is not necessary to disguise a well-designed poke at the ecological web by erecting a vague, general hypothesis to be tested. Careful, well-designed perturbations can be useful in producing basic information about an unknown system for which it is not yet possible to erect hypotheses leading to predictions of expected outcomes.

Motokawa (1989) argues that Western science is primarily hypothesis-oriented, whereas Eastern science (Japanese in particular) is more fact-oriented, a difference he attributes to different world views, which are paralleled in the different outlooks of Western religions and those of the East, Zen Buddhism in particular. I can find no such dichotomy in the field experiments with spiders performed by Japanese and Western scientists. I confess, though, that this restricted comparison is not a particularly powerful test of Motokawa's thesis. It would be helpful to recognize that the tension he describes exists within the Western scientific tradition. In contemporary ecology 'pure description' of a system is not considered as worthwhile as 'strict hypothesis testing.' The importance of pure description will become more widely recognized as community ecologists seek more mechanistic explanations of restricted sets of ecosystems. As we restrict the generality of our models, realizing that it is often folly to force generalizations across widely different types of ecosystems, the value of detailed descriptions will become more obvious. We should keep our outlook broad: well-planned field experiments can provide valuable revelations as well as yield the most direct tests of specific hypotheses.

A fantasy

Most of us who perform field experiments view the world in terms of replicable units of habitat that can be perturbed in order to test particular hypotheses about the larger system and, ideally, to test hypotheses that apply to similar systems elsewhere. Sometimes these hypotheses arise from knowledge of patterns derived from non-experimental studies. More often, though, broader issues, such as hypotheses about interspecific competition, motivate particular experiments. What would ecology gain if for one year experimentalists abandoned both approaches? Is it idle fantasy to consider this option? I have wondered at times what I would learn if I spent the entire field season sitting quietly in the forest where I conduct research, at all times of the day and night, just watching and listening. Of course I would miss a lot, because many organisms are secretive. I would have to change my focus periodically, sometimes examining the litter, other times observing the understorey vegetation and higher. By not running traps or extracting litter samples I would overlook much, and by not performing field experiments I would have no direct measure of the dynamics connecting components of the forest's ecological web. But how much do I miss by always entering the forest with ideas of hypotheses to test and timetables of quantitative data to collect?

Dangling threads and partially spun tales

Understudies and neglected actors

Experimentalists have given web-spinning spiders the most attention. The advantages they offer are obvious. Nevertheless, cursorial families are widespread and form a major component of the spider fauna; they deserve more study. Lycosids, conspicuous because of their abundance and activity patterns, have been the focus of much research, but very few field experiments have been conducted with wolf spiders. Families that are more secretive or that are active primarily at night have received practically no attention, either from experimentalists or from ecologists interested in documenting basic aspects of behavior and population ecology. Considerable challenges await researchers who attempt detailed experimental studies of these groups, but the rewards in terms of unexpected results could be large. These groups should be studied much more intensively. There is a danger that what we have discovered about the relatively small fraction of the cast that constitutes the collective spider persona will become entrenched as dogma about all spiders.

Gaps and tangled sectors

Despite the number and diversity of field experiments with spiders, what we know of their ecological webs is mostly gaps or tangled sectors. We need to spin a much stronger story. Below I offer suggestions of what I feel are particularly promising areas of research with spiders. In most of these selected research areas, investigators have already completed field experiments, many of them well designed. Nevertheless, more experimentation is needed to provide answers to many questions that, for the most part, we have just begun to address. These areas are the following:

Territoriality: Riechert has proposed that resource-based territoriality is widespread among spiders. If true, this would have major implications for the role of spiders in regulating prey populations, as well as revealing much about the evolution of agonistic behavior among generalist predators. Riechert's elegant, long-term research has clearly established that territoriality based upon prey availability is important in populations of the desert spider *Agelenopsis aperta*, but the existence of such a territorial system has not been established for other spiders. We need to know whether or not resource-based territoriality is widespread among spiders.

Optimal foraging theory: Several investigators have identified web builders and even a wandering spider (a thomisid) as suitable model systems for testing predictions of optimal foraging theory. Areas that seem fruitful to pursue are the influence on foraging behavior of (1) variance in prey availability, (2) exposure to natural enemies, and (3) exposure to abiotic disturbances.

Exploitative competition among wandering spiders: Practically all experimental studies of food limitation and exploitative competition (intra- and interspecific) have dealt with web-building spiders. The logistical challenges of experimenting with cursorial species have made web builders more attractive, but the non-web spiders should not be neglected any longer. Our generalizations about resource competition among spiders may be resting on a rather narrow base.

Limitation of prey populations: Here the experimental evidence is tantalizing but far from complete. The impact of spiders on prey populations has been rarely studied with the long-term, carefully controlled field experiments that are needed to account for effects of barriers and small plot size, and that will reveal the complexity of indirect effects expected to surface. This deficiency is particularly the case with natural ecosystems. The experiments with agroecosystems are more numerous, and in some instances better designed, but many more need to be done.

Intraguild predation (IGP): How does predation between spider species, and between spiders and other generalist arthropod predators, affect their densities and, as a consequence, their impact upon prey populations? Do spiders compete for prey with other generalist predators, or is IGP the primary interaction?

Natural enemies: Most field experiments with spiders have focused on trophic connections with their prey, yet spiders occupy an intermediate position in their food webs. Spiders are preyed upon not only by other members of the guild of generalist arthropod predators, but also by an array of natural enemies, invertebrate and vertebrate, from higher trophic levels. Several ecologists have used field experiments to assess the influence of natural enemies on spider populations. Continuing this approach will improve our understanding of how spiders connect to other components of their ecological webs.

Early instars: Ecologists understandably have avoided experimenting with the smaller stages, yet resource limitation, competition, and mortality from physical factors and natural enemies among early instars may be key factors in the population dynamics of many species. It is unlikely that ecologists will stampede to start research with stages that offer so many logistical challenges. The results of such a shift in research emphasis might be surprising.

Between-generation effects: Rarely have experiments examined the consequences of a perturbation, such as an increase in food supply or population density, upon densities of the target species in the next generation. Usually within-generation parameters are examined. More long-term experiments should be conducted with spiders. There is a need to continue pulse experiments for a longer duration, and also to conduct greater numbers of long-term press studies. For example: What are the consequences for the next generation of a doubled fecundity – does increased egg reproduction translate into more reproducing females next year? This question could be answered by following the population for another year without continuing to supplement food (a long-term pulse experiment). A supporting press experiment, in which prey supply was augmented for both hatchlings and older stages, would provide a helpful comparison. Of course such experiments are not easy to conduct, nor easy to interpret. For example, dispersal of the hatchlings could give a false indication of convergence in the next generation if the experimental units are smaller than the dispersal range of the young.

Untangling a tangled maze

Opportunities exist for diverse experimental research programs with spiders. Promising research directions need not be pursued exclusively of others; in fact, fruitful overlap is usually unavoidable. For example, detailed studies of territorial behavior, intraspecific competition and foraging behavior of major spiders in a community would aid in interpreting the results of long-term manipulations designed to uncover indirect effects involving spiders, IGP and prey populations. In addition, non-experimental investigations, such as characterizing spider diets in the field using biochemical techniques (Sunderland 1988), are indispensable in augmenting the experimental approach. Numerous other examples could be mentioned, each focusing on a different level of complexity in the ecological webs of spiders.

Results of field experiments have already shown that the spider persona is a multi-faceted character, with roles that differ markedly, either on the same stage or between different dramas. In terms of interactions with natural enemies and prey populations, web-spinning spiders of the forest understorey may have little in common with cursorial spiders that range across the vegetation. The persona of these web builders may resemble even less that of cursorial spiders in the leaf litter. Future experimental research with spiders should focus more on the differences between the spider characters, recognize the likelihood that separate plays have different scripts, and give increasing recognition to the importance of non-arachnid actors.

Future experimental research with spiders will be most productive if it directly attacks the complexity of ecological communities. Because non-experimental supporting research aids in interpreting perturbation studies of complex systems, field experiments with spiders will come to constitute one approach among several in research programs that will increasingly rely upon teams of investigators asking interwoven questions on the same community.

To be most productive, such integrated research programs require identification of appropriate communities. If field experimentation is to be an integral part of such efforts, some systems clearly are more suitable than others. Several come to mind: communities found in agroecosystems, on small islands, in salt marshes, in grasslands and on the forest floor. Results of previous research suggest that major findings will come from continued experimentation with the spiders in these ecological webs.

Synopsis

Field experiments have revealed several features of the spider persona. The spider in nature is usually hungry. Although food is frequently a limited resource, spiders do not often compete with other spiders for prey. Competition is not avoided because spiders have assumed different roles, but is absent because spider densities are below competitive levels. Experiments have uncovered several factors that account for the absence of strong exploitative competition in spider populations. No single explanation emerges; the ecological web of spiders is too intricate for one to expect a simple pattern. Portions of the spider persona remain enigmatic. For example, convincing arguments lead to the conclusion that spiders should not regulate prey populations, yet other indirect evidence suggests that the entire collection of spiders in a community should cause substantial density-independent control of insect populations. These portraits are not necessarily contradictory, but they highlight the dangers inherent in oversimplification.

Clearly the spider persona is complicated; no single ecological role for the spider can be postulated. In fact, future research should reveal numerous different portraits, each drawn largely along family lines. The major goal of experimental research on spiders, however, ought not to be aimed at fleshing out a more detailed characterization of a model organism. We should shift efforts away from trying to describe one type of model predator, to characterizing the interactions occurring within communities that are representative of a wide class of systems. One important criterion in selecting a model system is its amenability to field experimentation.

Spiders, particularly species that spin webs, offer many advantages to the experimentalist. Field experiments have modified some preconceptions ecologists have held about spiders. Experiments have shown that food shortages act as a density-independent factor for many web-building populations. Failure to find evidence of competition in several field manipulations has weakened competition-based explanations of spider communities. Field experiments in agroecosystems are forcing ecologists to re-think the possible roles of spiders as biocontrol agents.

Field experiments with spiders have not only produced counter-intuitive results; they have also confirmed hypotheses about spiders (e.g. the importance of food limitation). Field experiments have confirmed the hypothesis that lizards affect the abundance of spiders, a prediction based upon an observed negative correlation between lizard and spider numbers on a series of small Bahamian islands. This latter type of correlation, usually termed a *natural experiment*, is a valuable source of hypotheses, but is qualitatively different from the planned field experiment. The label of *experiment* should be reserved for those studies in which the investigator alters variables in a previously unmanipulated set of experimental units, according to accepted principles of experimental design.

Field experiments with spiders highlight problems of design and interpretation that are central to all field experimentation in ecology. The major issues condense into (1) meeting the challenge of establishing sufficient numbers of replicates while maintaining plot size large enough to include a reasonably intact portion of the system under study, and (2) establishing controls for the procedures used to achieve the manipulation, e.g. controlling for 'fence effects.'

Many ecologists have been reluctant to recognize the power of *negative results*, a term erroneously applied to outcomes of experiments in which differences between treatment and control plots are statistically insignificant. Properly planned and executed experiments that fail to support a hypothesis provide valid evidence against the hypothesis. Eventual acceptance or rejection of the hypothesis depends upon the pattern of results from several experiments. For example: solid evidence for interspecific competition between web-spinning spiders has been found in one field experiment, but several others have failed to uncover competition. Unless contradictory evidence is collected in the future, we should reject the hypothesis that interspecific competition is a major interaction among web-building spiders. Experiments that produce statistically insignificant results will make even more positive contributions to testing and developing theory if the investigators are more conscious of the statistical power of the tests they employ.

Field experiments are valuable not only as tests of specific hypotheses; they also can expand basic knowledge of a system by uncovering unexpected responses to perturbations. Detection of indirect effects is perhaps the most obvious example. As ecologists increasingly turn their attention to experimentation with complex systems, field experiments should become viewed as a legitimate means for uncovering basic information about a system, in addition to serving as a vehicle for testing carefully specified hypotheses.

Many fertile areas of research with spiders remain to be explored with research programs that emphasize field experimentation. Some examples of topics that merit further research with spiders are *territoriality, optimal foraging theory, exploitative competition among wandering spiders, prey limitation in natural communities and agroecosystems, intraguild predation, impact of natural enemies, ecology of early instars, and between-generation effects.*

Spiders are integral to the ecological webs of most terrestrial ecosystems. Future experimental research with spiders has the potential to make major contributions if it directly attacks the complexity of ecological communities. Based upon experiments conducted so far, promising model systems include communities found in agroecosystems, on small islands, in salt marshes, in grasslands and on the forest floor.

References

Aart, P. J. M., van der, & T. de Wit. 1971. A field study on interspecific competition between ants (Formicidae) and hunting spiders (Lycosidae, Gnaphosidae, Ctenidae, Pisauridae, Clubionidae). *Netherlands Journal of Zoology* 21:117–126.

Abrams, P. A. 1983. The theory of limiting similarity. *Annual Review of Ecology and Systematics* 14:359–376.

Abrams, P. A. 1986. Character displacement and niche shift analyzed using consumer-resource models of competition. *Theoretical Population Biology* 29:107–160.

Abrams, P. A. 1987a. Indirect interactions between species that share a predator: varieties of indirect effects. In: *Predation: Direct and Indirect Impacts on Aquatic Communities*, W. G. Kerfoot & A. Sih, eds., pp. 38–54. Hanover, NH, University Press of New England.

Abrams, P. A. 1987b. On classifying interactions between populations. *Oecologia* 73:272–281.

Abrams, P. A. 1990. Ecological vs. evolutionary consequences of competition. *Oikos* 57:147–151.

Addo, P. E. A. 1968. *Ghana Folk Tales: Ananse Stories from Africa*. New York, Exposition Press.

Aitchison, C. W. 1984. A possible subnivean food chain. *Carnegie Museum of Natural History, Special Publication* 10:363–372.

Aitchison, C. W. 1987. Feeding ecology of winter-active spiders. In: *Ecophysiology of Spiders*, W. Nentwig, ed., pp. 264–273. Berlin, New York, London, Paris and Tokyo, Springer.

Albert, A. M. 1976. Biomasse von Chilopoden in einem Buchenaltbestand des Solling. *Verhandlungen der Gesellschaft für Ökologie* 6: 93–101.

Anderson, J. F. 1970. Metabolic rates of spiders. *Comparative Biochemistry and Physiology* 33:51–72.

Anderson, J. F. 1974. Responses to starvation in the spiders *Lycosa lenta* Hentz and *Filistata hibernalis* (Hentz). *Ecology* 55:576–585.

Anderson, J. F. & K. N. Prestwich. 1982. Respiratory gas exchange in spiders. *Physiological Zoology* 55:72–90.

Andrewartha, H. G. & L. C. Birch. 1954. *The Distribution and Abundance of Animals*. Chicago, University of Chicago Press.

Andrewartha, H. G. & L. C. Birch. 1984. *The Ecological Web*. Chicago, University of Chicago Press.

Andrzejewska, L. 1971. Productivity investigations of two types of meadows in the

Vistula Valley. VI. Production and population density of leafhopper (Homoptera–Auchenorrhyncha) communities. *Ekologia Polska* 19:151–172.

Andrzejewska, L., A. Breymeyer, A. Kajak & Z. Wojcik. 1967. Experimental studies on trophic relationships of terrestrial invertebrates. In: *Secondary Productivity of Terrestrial Ecosystems*, K. Petrusewicz, ed., pp. 477–495. Warsaw, Panstwowe Wydawnictwo Naukowe.

Archer, A. F. 1946. The Theridiidae or comb-footed spiders of Alabama. *Alabama Museum of Natural History Paper* 22:1–67.

Askenmo, C., A. von Broemssen, J. Eckman & C. Jansson. 1977. Impact of some wintering birds on spider abundance in spruce. *Oikos* 28:90–94.

Atchley, W. R., C. T. Gaskins & D. Anderson. 1976. Statistical properties of ratios. I. Empirical results. *Systematic Zoology* 25:137–148.

Barnard, C.J. (ed.). 1984. *Producers and Scroungers: Strategies of Exploitation and Parasitism*. Beckenham, Croom Helm.

Barnes, R. D. 1953. The ecological distribution of spiders in nonforest maritime communitites at Beaufort, North Carolina. *Ecological Monographs* 23:315–337.

Barnes, R. D. & B. M. Barnes. 1955. The spider population of the abstract broomsedge community of the southeastern Piedmont. *Ecology* 36:658–666.

Barth, F. (ed.). 1985. *Neurobiology of Arachnids*. Berlin, New York, London, Paris and Tokyo, Springer.

Begon, M., J. L. Harper & C. R. Townsend. 1990. *Ecology: Individuals, Populations, and Communities*. Oxford, Boston and Melbourne, Blackwell Scientific Publications.

Bender, E. A., T. J. Case & M. E. Gilpin. 1984. Perturbation experiments in community ecology: theory and practice. *Ecology* 65:1–13.

Biere, J. M. & G. W. Uetz. 1981. Web orientation in the spider *Micrathena gracilis* (Walckenaer) (Araneae: Araneidae). *Ecology* 62:336–344.

Bishop, A. L. & P. R. B. Blood. 1981. Interactions between natural populations of spiders and pests in cotton and their importance to cotton production in Southeastern Queensland. *General and Applied Entomology* 13:98–104.

Bleckmann, H. & J. Rovner. 1984. Sensory ecology of a semi-aquatic spider (*Dolomedes triton*). *Behavioral Ecology and Sociobiology* 14:297–301.

Bradley, R. A. 1983. Complex food webs and manipulative experiments in ecology. *Oikos* 41:150–152.

Breymeyer, A. 1966. Relations between wandering spiders and other epigeic predatory *Arthropoda*. *Ekologia Polska – Seria A* 14:27–71 (cited by van der Aart & De Wit 1971).

Breymeyer, A. 1967. Preliminary data for estimating the biological production of wandering spiders. In: *Secondary Productivity in Terrestrial Ecosystems*, Vol. 2, K. Petrusewicz, ed., pp. 821–834. Warsaw, Panstwowe Wydawnictwo Naukowe.

Bristowe, W. S. 1939,1941. *The Comity of Spiders*. 2 vols. London, The Ray Society. Republished 1968 by Johnson Reprint Corp., New York.

Bristowe, W. S. 1971. *The World of Spiders*. London, Collins.

Brown, J. H., D. W. Davidson, J. C. Munger & R. S. Inouye. 1986. Experimental community ecology: the desert granivore system. In: *Community Ecology*, J. Diamond & T. J. Case, eds., pp. 41–62. New York, Harper and Row.

Brown, K. M. 1981. Foraging ecology and niche partitioning in orb-weaving spiders. *Oecologia* 50:380–385.

Bultman, T. L. & G. W. Uetz. 1982. Abundance and community structure of forest floor spiders following litter manipulation. *Oecologia* 55:34–41.

Bultman, T. L. & G. W. Uetz. 1984. Effect of structure and nutritional quality of litter on abundances of litter-dwelling arthropods. *American Midland Naturalist* 111:165–172.

Bultman, T. L., G. W. Uetz & A. R. Brady. 1982. A comparison of cursorial spider communities along a successional gradient. *Journal of Arachnology* 10:23–35.

Burgess, J. W. 1976. Social spiders. *Scientific American* 234(3):101–106.

Buskirk, R. E. 1981. Sociality in the Arachnida. In: *Social Insects*, Vol. 2, H. R. Hermann, ed., pp. 281–367. New York, Academic Press.

Cady, A. B. 1983. Microhabitat selection and locomotor activity of *Schizocosa ocreata* (Walckenaer) (Araneae: Lycosidae). *Journal of Arachnology* 11:297–307.

Caraco, T. & R.G. Gillespie. 1986. Risk sensitivity: foraging mode in an ambush predator. *Ecology* 67:1180–1185.

Castillo, J. A. & W. G. Eberhard. 1983. Use of artificial webs to determine prey available to orb weaving spiders. *Ecology* 64:1655–1658.

Cherrett, J. M. 1964. The distribution of spiders on the Moor House National Nature Reserve, Westmorland. *Journal of Animal Ecology* 33:27–48.

Chew, R. M. 1961. Ecology of the spiders of a desert community. *Journal of the New York Entomological Society* 69: 5–41.

Chew, R. M. 1974. Consumers as regulators of ecosystems: an alternative to energetics. *Ohio Journal of Science* 74:359–371.

Chiverton, P. A. 1986. Predator density manipulation and its effects on populations of *Rhopalosiphum padi* (Hom.: Aphididae) in spring barley. *Annals of Applied Biology* 109:49–60.

Christenson, T. E. 1984. Alternative reproductive tactics in spiders. *American Zoologist* 24:321–332.

Clarke, R. D. & P. R. Grant. 1968. An experimental study of the role of spiders as predators in a forest litter community. Part I. *Ecology* 49:1152–1154.

Cody, M. L. 1974. *Competition and the Structure of Bird Communities*. Princeton, Princeton University Press.

Cohen, J. 1977. *Statistical Power Analysis for the Behavioral Sciences*. New York, Academic Press.

Colebourn, P. H. 1974. The influence of habitat structure on the distribution of *Araneus diadematus* Clerck. *Journal of Animal Ecology* 43:401–409.

Colwell, R. K. & D. J. Futuyma. 1971. On the measurement of niche breadth and overlap. *Ecology* 52:567–576.

Conley, M. R. 1985. Predation versus resource limitation in survival of adult burrowing wolf spiders (Araneae, Lycosidae). *Oecologia* 67:71–75.

Connell, J. H. 1975. Some mechanisms producing structure in natural communities. In: *Ecology and Evolution of Communities*, M. L. Cody & J. M. Diamond, eds., pp. 460–490, Cambridge, Mass., Harvard University Press.

Connell, J. H. 1980. Diversity and the coevolution of competitors, or the ghost of competition past. *Oikos* 35:131–138.

Connell, J. H. 1983. On the prevalence and relative importance of interspecific competition: evidence from field experiments. *American Naturalist* 122:661–696.

Connell, J. H. 1985. On testing models of competitive coevolution. *Oikos* 45:298–300.

Corrigan, J. E. & R. G. Bennett. 1987. Predation by *Chiracanthium mildei* (Araneae, Clubionidae) on larval *Phyllonorycter blancardella* (Lepidoptera, Gracillariidae) in a greenhouse. *Journal of Arachnology* 15:132–134.

Coulson, J. C. & J. Butterfield. 1986. The spider communities on peat and upland grasslands in northern England. *Holarctic Ecology* 9:229–239.

Coville, R. E. 1987. Spider–hunting sphecid wasps. In: *Ecophysiology of Spiders*, W. Nentwig, ed., pp. 309–318. Berlin, New York, London, Paris and Tokyo, Springer.

Craig, C. L., A. Okubo & V. Andreasen. 1985. Effect of spider orb–web and insect oscillations on prey interception. *Journal of Theoretical Biology* 115:201–211.

Crawford, R. L. & J. S. Edwards. 1986. Ballooning spiders as a component of arthropod fallout on snowfields of Mount Rainier, Washington, U.S.A. *Arctic and Alpine Research* 18:429–437.

Culin, J. D. & K. V. Yeargan. 1983. Comparative study of spider communities in alfalfa and soybean ecosystems: foliage-dwelling spiders. *Annals of the Entomological Society of America* 76:825–831.

Dabrowska-Prot, E. & J. Luczak. 1968a. Spiders and mosquitoes of the ecotone of alder forest (*Carici Elongatae–Alnetum*) and oak-pine forest (*Pino-Quercetum*). *Ekologia Polska–Seria A* 16:461–483.

Dabrowska-Prot, E. & J. Luczak. 1968b. Studies on the incidence of mosquitoes in the food of *Tetragnatha montana* Simon and its food activity in the natural habitat. *Ekologia Polska–Seria A* 16:843–852.

Dabrowska-Prot, E., J. Luczak & K. Tarwid. 1968. Prey and predator density and their reactions in the process of mosquito reduction by spiders in field experiments. *Ekologia Polska–Seria A* 16:773–819.

Dabrowska-Prot, E., J. Luczak & Z. Wojcik. 1973. Ecological analysis of two invertebrate groups in the wet alder wood and meadow ecotone. *Ekologia Polska* 21:753–812.

Dahl, F. 1906. Die physiologische Zuchtwahl im weiteren Sinne. *Biologisches Zentralblatt* 26:3–15 (cited by Tretzel 1955).

Darwin, C. 1859. *The Origin of Species*. London, John Murray.

Davidson, D. W. 1980. Some consequences of diffuse competition in a desert ant community. *American Naturalist* 116:92–105.

Dean, D. A., W. L. Sterling, M. Nyffeler & R. G. Breene. 1987. Foraging by selected spider predators on the cotton fleahopper and other prey. *Southwestern Entomologist* 12:263–270.

Decae, A. E. 1987. Dispersal: ballooning and other mechanisms. In: *Ecophysiology of Spiders*, W. Nentwig, ed., pp. 348–356. Berlin, New York, London, Paris and Tokyo, Springer.

Diamond, J. 1986. Overview: laboratory experiments, field experiments, and natural experiments. In: *Community Ecology*, J. Diamond & T. J. Case, eds., pp. 3–22. New York, Harper and Row.

Döbel, H. G. 1987. The role of spiders in the regulation of salt marsh planthopper populations. Unpubl. Master's Thesis, University of Maryland.

Döbel, H. G., R. F. Denno & J. A. Coddington. 1990. Spider (Araneae) community structure in an intertidal salt marsh: effects of vegetation structure and tidal flooding. *Environmental Entomology* 19:1356–1370.

Dobzhansky, T. 1950. Evolution in the tropics. *American Scientist* 38:209–221.

Dondale, C. D. 1977. Life histories and distribution patterns of hunting spiders (Araneida) in an Ontario meadow. *Journal of Arachnology* 4:73–93.

Dondale, C. D. & M. R. Binns. 1977. Effect of weather factors on spiders (Araneida) in an Ontario meadow. *Canadian Journal of Zoology* 55:1336–1341.

Drift, J. van der. 1951. Analysis of the animal community in a beech forest floor. *Tijdschrift voor Entomologie* 94:1–168.

Duffey, E. 1975. Habitat selection by spiders in man-made environments. In: *Proceedings of the Sixth International Arachnological Congress*, pp. 53–67. Amsterdam.

Duffey, E. 1978. Ecological strategies in spiders including some characteristics of species in pioneer and mature habitats. *Symposium of the Zoological Society of London* 42:109–123.

Eberhard, W. G. 1971. The ecology of the web of *Uloborus diversus* (Araneae: Uloboridae). *Oecologia* 6:328–342.

Eberhard, W. G. 1980. The natural history and behavior of the bolas spider *Mastophora dizzydeani* sp. n. (Araneae). *Psyche* 87:143–169.

Eberhard, W. G. 1989. Niche expansion in the spider *Wendilgarda galapagensis* (Araneae, Theridiosomatidae) on Cocos Island. *Revista de Biología Tropical* 37:163–168.

Eberhard, W. G. 1990. Function and phylogeny of spider webs. *Annual Review of Ecology and Systematics* 21:341–372.

Eberhardt, L. L. & J. M. Thomas. 1991. Designing environmental field studies. *Ecological Monographs* 61:53–73.

Edgar, W. D. 1969. Prey and predators of the wolf spider *Lycosa lubugris*. *Journal of Zoology, London* 159:405–411.

Edgar, W. D. 1971. The life-cycle, abundance and seasonal movement of the wolf spider, *Lycosa (Pardosa) lugubris*, in central Scotland. *Journal of Animal Ecology* 40:303–322.

Eisner, T. & S. Nowicki. 1983. Spider web protection through visual advertisement: role of the stabilimentum. *Science* 219:185–187.

Enders, F. 1974. Vertical stratification of orb-web spiders and a consideration of other methods of coexistence. *Ecology* 55:317–328.

Enders, F. 1975a. Effects of prey capture, web destruction and habitat physiognomy in web–site tenacity of *Argiope* spiders (Araneidae). *Journal of Arachnology* 3:75–82.

Enders, F. 1975b. The influence of hunting manner on prey size, particularly in spiders with long attack distances (Araneae, Linyphiidae and Salticidae). *American Naturalist* 109:737–763.

Enright, J. T. 1976. Climate and population regulation: the biogeographer's dilemma. *Oecologia* 24:295–310.

Exline, H. & H. W. Levi. 1962. American spiders of the genus *Argyrodes* (Araneae, Theridiidae). *Bulletin of the Harvard Museum of Comparative Zoology* 127:75–204.

Fincke, O. M., L. Higgins & E. Rojas. 1990. Parasitism of *Nephila clavipes* (Araneae, Tetragnathidae) by an ichneumonid (Hymenoptera, Polysphinctini) in Panama.

Journal of Arachnology 18:321–329.

Fink, L. S. 1986. Costs and benefits of maternal behaviour in the green lynx spider (Oxyopidae, *Peucetia viridans*). *Animal Behaviour* 34:1051–1060.

Fink, L. S. 1987. Green lynx spider egg sacs: sources of mortality and the function of female guarding (Araneae, Oxyopidae). *Journal of Arachnology* 15:231–239.

Fitch, H. A. 1963. Spiders of the University of Kansas Natural History Reservation and Rockefeller Experimental Tract. *University of Kansas Museum of Natural History Museum Miscellaneous Publication No. 33.*

Foelix, R. 1982. *Biology of Spiders.* Cambridge, Mass., Harvard University Press.

Fowler, H. G. & W. G. Whitford. 1985. Structure and organization of a winter community of cavity-inhabiting, web-building spiders (Pholcidae and Theridiidae) in a Chihuahuan Desert habitat. *Journal of Arid Environments* 8:57–65.

Fretwell, S.D. 1987. Food chain dynamics: the central theory of ecology? *Oikos* 50:291–301.

Fritz, R. S. & D. H. Morse. 1985. Reproductive success and foraging of the crab spider *Misumena vatia*. *Oecologia* 65:194–200.

Furuta, K. 1977. Evaluation of spiders, *Oxyopes sertatus* and *O. badius* (Oxyopidae) as a mortality factor of gypsy moth, *Lymantria dispar* (Lepidoptera: Lymantriidae) and pine moth, *Dendrolimus spectabilis* (Lepidoptera: Lasiocampidae). *Applied Entomology and Zoology* 12:313–324.

Gasdorf, E. C. & C. J. Goodnight. 1963. Studies on the ecology of soil arachnids. *Ecology* 44:261–268.

Gertsch, W. J. 1979. *American Spiders.* New York, Van Nostrand Reinhold.

Gertsch, W. J. & S. E. Riechert. 1976. The spatial and temporal partitioning of a desert spider community, with descriptions of new species. *Novitates (American Museum)* 2604:1–25.

Gillespie, R. G. 1981. The quest for prey by the web-building spider *Amaurobius similis* (Blackwell). *Animal Behaviour* 35:675–681.

Gillespie, R. G. 1987. The mechanism of habitat selection in the long-jawed orb-weaving spider *Tetragnatha elongata* (Araneae, Tetragnathidae). *Journal of Arachnology* 15:81–90.

Gillespie, R. G. & T. Caraco. 1987. Risk-sensitive foraging strategies of two spider populations. *Ecology* 68:887–899.

Gilpin, M. E., M. P. Carpenter & M. J. Pomerantz. 1986. The assembly of a laboratory community: multispecies competition. In: *Community Ecology*, J. Diamond & T. J. Case, eds., pp. 23–40. New York, Harper and Row.

Givens, R. P. 1978. Dimorphic foraging strategies of a salticid spider (*Phidippus audax*). *Ecology* 59:309–321.

Greene, E., L. J. Orsak & D. W. Whitman. 1987. A tephritid fly mimics the territorial displays of its jumping spider predators. *Science* 236:310–312.

Greenstone, M. H. 1978. The numerical response to prey availability of *Pardosa ramulosa* (McCook) (Araneae: Lycosidae) and its relationship to the role of spiders in the balance of nature. *Symposia of the Zoological Society of London* 42:183–193.

Greenstone, M. H. 1979. Spider feeding behaviour optimises dietary essential amino acid composition. *Nature* 282:501–503.

Greenstone, M. H. 1980. Contiguous allotopy of *Pardosa ramulosa* and *Pardosa tuoba* (Araneae: Lycosidae) in the San Francisco Bay Region, and its implications for patterns of resource partitioning in the genus. *American Midland Naturalist* 104:305–311.

Greenstone, M. H. 1984. Determinants of web spider species diversity: vegetation structural diversity vs. prey availability. *Oecologia* 62:299–304.

Greenstone, M. H. & A. F. Bennett. 1980. Foraging strategy and metabolic rate in spiders. *Ecology* 61:1255–1259.

Greenstone, M. H., C. E. Morgan & A. L. Hultsch. 1987. Ballooning spiders in Missouri, USA, and New South Wales, Australia: family and mass distributions. *Journal of Arachnology* 15:163–170.

Gunnarsson, B. 1983. Winter mortality of spruce-living spiders: effect of spider interactions and bird predation. *Oikos* 40:226–233.

Gunnarsson, B. 1985. Interspecific predation as a mortality factor among overwintering spiders. *Oecologia* 65:498–502.

Gunnarsson, B. 1987. Sex ratio in the spider *Pityohyphantes phrygianus* affected by winter severity. *Journal of Zoology (London)* 213:609–619.

Gunnarsson, B. 1988. Spruce-living spiders and forest decline: the importance of needle-loss. *Biological Conservation* 43:309–319.

Gunnarsson, B. 1990. Vegetation structure and the abundance and size distribution of spruce-living spiders. *Journal of Animal Ecology* 59:743–752.

Hagstrum, D. W. 1970. Ecological energetics of the spider *Tarentula kochi*. *Annals of the Entomological Society of America* 63:1297–1304.

Hairston, N. G. 1989. *Ecological Experiments: Purpose, Design and Execution*. Cambridge, New York and Melbourne, Cambridge University Press.

Hairston, N. G., F. E. Smith & L. B. Slobodkin. 1960. Community structure, population control, and competition. *American Naturalist* 94:421–425.

Hallander, H. 1970a. Prey, cannibalism, and microhabitat selection in the wolf spiders *Pardosa chelata* O. F. Muller and *P. pullata* Clerck. *Oikos* 21:337–340.

Hallander, H. 1970b. Environments of the wolf spiders *Pardosa chelata* (O. F. Muller) and *Pardosa pullata* (Clerck). *Ekologia Polska* 18:43–72.

Hardman, J. M. & A. L. Turnbull. 1974. The interaction of spatial heterogeneity, predator competition and the functional response to prey density in a laboratory system of wolf spiders (Araneae: Lycosidae) and fruit flies (Diptera: Drosophilidae). *Journal of Animal Ecology* 43:155–171.

Hardman, J. M. & A. L. Turnbull. 1980. Functional response of the wolf spider, *Pardosa vancouveri*, to changes in the density of vestigial-winged fruit flies. *Researches on Population Ecology* 21:233–259.

Hatley, C. L. & J. A. MacMahon. 1980. Spider community organization: seasonal variation and the role of vegetation architecture. *Environmental Entomology* 9:632–639.

Haynes, D. L. & P. Sisojevic. 1966. Predatory behavior of *Philodromus rufus* Walckenaer (Araneae: Thomisidae). *Canadian Entomologist* 98:113–133.

Heidger, C. 1988. Ecology of spiders inhabiting abandoned mammal burrows in South African savanna. *Oecologia* 76:303–306.

Hodge, M. A. 1987a. Macrohabitat selection by the orb weaving spider, *Micrathena gracilis*. *Psyche* 94:347–361.

Hodge, M. A. 1987b. Factors influencing web site residence time of the orb weaving spider, *Micrathena gracilis*. *Psyche* 94:363–371.

Hoffmaster, D. K. 1985a. Resource breadth in orb-weaving spiders: a tropical–temperate comparison. *Ecology* 66:626–629.

Hoffmaster, D. K. 1985b. Community composition and nearest neighbor relationships in orb-weaving spiders: the product of aggression? *Behavioral Ecology and Sociobiology* 16:349–353.

Hoffmaster, D. K. 1986. Aggression in tropical orb-weaving spiders: a quest for food? *Ethology* 72:265–276.

Hogstad, O. 1984. Variation in numbers, territoriality and flock size of a Goldcrest *Regulus regulus* population in winter. *Ibis* 126:296–306.

Hollander, J. den & H. Lof. 1972. Differential use of the habitat by *Pardosa pullata* (Clerck) and *Pardosa prativaga* (L. Koch) in a mixed population. *Tijdschrift voor Entomologie* 115:205–215.

Holling, C. S. 1959. The components of predation as revealed by a study of small-mammal predation of the European sawfly. *Canadian Entomologist* 91:293–320.

Holt, R. D. 1977. Predation, apparent competition and the structure of prey communities. *Theoretical Population Biology* 12:197–229.

Horton, C. C. 1980. A defensive function for the stabilimenta of two orb weaving spiders (Araneae, Araneidae). *Psyche* 87:13–20.

Horton, C. C. & D. H. Wise. 1983. The experimental analysis of competition between two syntopic species of orb-web spiders (Araneae: Araneidae). *Ecology* 64:929–944.

Hubbell, T. H. 1932. An unusual occurrence of spiders in northern Florida. *Annals of the Entomological Society of America* 25:502–504.

Huey, R. B., E. R. Pianka & T. W. Schoener. 1983. *Lizard Ecology: Studies of a Model Organism.* Cambridge, Mass., Harvard University Press.

Huhta, V. 1971. Succession in the spider communities of the forest floor after clear-cutting and prescribed burning. *Annales Zoologici Fennici* 8:483–542 (cited by Bultman *et al.* 1982).

Huhta, V. & J. Viramo. 1979. Spiders active on snow in northern Finland. *Annales Zoologici Fennici* 16:169–176.

Hurd, L. E. & R. M. Eisenberg. 1990. Arthropod community responses to manipulation of a bitrophic predator guild. *Ecology* 71: 2107–2114.

Hurlbert, S. H. 1984. Pseudoreplication and the design of ecological field experiments. *Ecological Monographs* 54:187–211.

Hutchinson, G. E. 1959. Homage to Santa Rosalia, or why are there so many kinds of animals? *American Naturalist* 93:145–159.

Hutchinson, G. E. 1965. *The Ecological Theater and the Evolutionary Play.* New Haven, Conn., Yale University Press.

Itô, Y. 1964. Preliminary studies on the respiratory energy loss of a spider, *Lycosa pseudoannulata. Researches on Population Ecology* 6:13–21.

Jackson, R. R. 1988. The biology of *Jacksonoides queenslandica*, a jumping spider (Araneae: Salticidae) from Queensland: intraspecific interactions, web-invasion, predators, and prey. *New Zealand Journal of Zoology* 15:1–37.

Jackson, R. R. & S. E. Hallas. 1986a. Predatory versatility and intraspecific interactions of spartateine jumping spiders (Araneae: Salticidae): *Brettus adonis, B. cingulatus, Cyrba algerina* and *Phaeacius* sp. indet. *New Zealand Journal of Zoology* 13:491–520.

Jackson, R. R. & S. E. Hallas. 1986b. Comparative biology of *Portia africana, P. albimana, P. fimbriata, P. labiata,* and *P. shultzi,* araneophagic, web-building jumping spiders (Araneae: Salticidae): utilisation of webs, predatory versatility, and intraspecific interactions. *New Zealand Journal of Zoology* 13:423–489.

Jacques, A. R. & L. M. Dill. 1980. Zebra spiders may use uncorrelated asymmetries to settle contests. *American Naturalist* 116:899–901.

Janetos, A. C. 1982a. Active foragers vs. sit-and-wait predators: a simple model. *Journal of Theoretical Biology* 95:381–385.

Janetos, A. C. 1982b. Foraging tactics of two guilds of web-spinning spiders. *Behavioral Ecology and Sociobiology* 10:19–27.

Janetos, A. C. 1986. Web-site selection: are we asking the right questions? In: *Spiders – Webs, Behavior and Evolution,* W. A. Shear, ed., pp. 9–22. Stanford, Calif., Stanford University Press.

Jansson, C. & A. von Brömssen. 1981. Winter decline of spiders and insects in spruce *Picea abies* and its relation to predation by birds. *Holarctic Ecology* 4:82–93.

Jocqué, R. 1973. The spider fauna of adjacent woodland areas with different humus types. *Biologisch Jaarboek* 41:153–179.

Jocqué, R. 1984. Considérations concernant l'abondance relative des araignées errantes et des araignées à toile vivant au niveau du sol. *Revue Arachnologique* 5:193–204.

Jones, R. L. 1981. *Report of the USDA Biological Control of Stem Borers Study Team's visit to the People's Republic of China, July 1980.* International Agricultural Programs Office, College of Agriculture, University of Minnesota, Minneapolis.

Kajak, A. 1965. An analysis of food relations between the spiders – *Araneus cornutus* Clerck and *Araneus quadratus* Clerck – and their prey in meadows. *Ekologia Polska–Seria A* 13:717–762.

Kajak, A. 1967. Productivity of some populations of web spiders. In: *Secondary Productivity of Terrestrial Ecosystems,* K. Petrusewicz, ed., pp. 807–820. Warsaw, Panstwowe Wydawnictwo Naukowe.

Kajak, A. 1971. Productivity investigation of two types of meadow in the Vistula Valley. *Ekologia Polska* 19:197–211.

Kajak, A. 1978a. Analysis of consumption by spiders under laboratory and field conditions. *Ekologia Polska* 26:409–427.

Kajak, A. 1978b. The effect of fertilizers on numbers and biomass of spiders in a meadow. *Symposium of the Zoological Society of London* 42:125–129.

Kajak, A. 1980a. Do the changes caused in spider communities by the application of fertilizers advance with time? In: *Proceedings of the Eighth International Congress of Arachnology,* pp. 115–119. Vienna, H. Egermann.

Kajak, A. 1980b. Invertebrate predator subsystem. In: *Grasslands, Systems Analysis and Man,* A. Breymeyer & G. M. Van Dyne, eds., pp. 539–589. Cambridge, New York and Melbourne, Cambridge University Press.

Kajak, A. 1981. Analysis of the effect of mineral fertilization on the meadow spider community. *Ekologia Polska* 29:313–326.

Kajak, A. & H. Jakubczyk. 1975. Experimental studies on spider predation. In: *Proceedings of the Sixth International Arachnological Congress,* pp. 82–85. Amsterdam.

Kajak, A. & H. Jakubczyk. 1976. The effect of intensive fertilization on the structure and productivity of meadow ecosystems. 18. Trophic relationships of epigeic predators. *Polish Ecological Studies* 2:219–229.

Kajak, A. & H. Jakubczyk. 1977. Experimental studies of predation in the soil-litter interface. In: *Soil Organisms as Components of Ecosystems*, U. Lohm & T. Persson, eds. *Proceedings of the 6th International Colloquium on Soil Zoology, Ecological Bulletins (Stockholm)* 25:493–496.

Kajak, A. & M. Kaczmarek. 1988. Effect of predation on soil mesofauna: an experimental study. *XI. Europäisches Arachnologisches Colloquium, Technische Universität Berlin, Dokumentation Kongresse und Tagungen* 38:188–198.

Kaston, B. J. 1965. Some little known aspects of spider behavior. *American Midland Naturalist* 73:336–356.

Kaston, B. J. 1978. *How to Know the Spiders*. Dubuque, Iowa, Wm. C. Brown Co. Publishers.

Kayashima, I. 1961. Study on the lynx-spider *Oxyopes sertatus* L. Koch for biological control of the Cryptomerian leaf-fly *Contarinia inouyei* Mani. *Review of Applied Entomology* 51:413 (an abstract of Kayashima, I. 1961. *Publication of the Entomological Laboratory of the College of Agriculture, University of Osaka* 6:167–169).

Kempson, D., M. Lloyd & R. Ghelardi. 1963. A new extractor for woodland litter. *Pedobiologia* 3:1–21.

Kenmore, P. E., F. O. Cariño, C. A. Perez, V. A. Dyck & A. P. Guttierrez. 1984. Population regulation of the rice brown planthopper (*Nilaparvata lugens* Stål) within rice fields in the Philippines. *Journal of Plant Protection in the Tropics* 1:19–37.

Kessler, A. 1971. Relation between egg production and food consumption in species of the genus *Pardosa* (Lycosidae, Araneae) under experimental conditions of food-abundance and food-shortage. *Oecologia* 8:93–109.

Kessler, A. 1973. A comparative study of the production of eggs in eight *Pardosa* species in the field (Araneae, Lycosidae). *Tijdschrift voor Entomologie* 116:23–41.

Kessler, A., J. W. C. Vermeulen & P. Wapenaar. 1984. Partitioning of the space in tussocks of the sedge, *Carex distans*, during winter by a spider community. *Journal of Zoology* 204:259–270.

Kirby, W. & W. Spence. 1815. *An Introduction to Entomology*. Vol. 1. London (cited by Cherrett 1964).

Kirchner, W. 1987. Behavioural and physiological adaptations to cold. In: *Ecophysiology of Spiders*, W. Nentwig, ed., pp. 66–77. Berlin, New York, London, Paris and Tokyo, Springer.

Kiritani, K. & N. Kakiya. 1975. An analysis of the predator-prey system in the paddy field. *Researches on Population Ecology* 17:29–38.

Kiritani, K., S. Kawahara, T. Sasaba & F. Nakasuji. 1972. Quantitative evaluation of predation by spiders on the green rice leafhopper, *Nephotettix cincticeps* Uhler, by a sight-count method. *Researches on Population Ecology* 13:187–200.

Kobayashi, S. 1975. The effect of *Drosophila* release on the spider population in a paddy field. *Applied Entomology and Zoology* 10:268–274.

Kronk, A. E. & S. E. Riechert. 1979. Parameters affecting the habitat choice of a desert wolf spider, *Lycosa santrita* Chamberlin and Ivie. *Journal of Arachnology* 7:155–166.

Kuenzler, E. J. 1958. Niche relations of three species of lycosid spiders. *Ecology*

39:494–500.

Kuhn, T. S. 1962. *The Structure of Scientific Revolutions.* Chicago, University of Chicago Press.

Kuhn, T. S. 1974. Second thoughts on paradigms. In: *The Structure of Scientific Theories,* F. Suppe, ed., pp. 459–482. Urbana, University of Illinois Press.

Kullmann, E. 1972. Evolution of social behavior in spiders (Araneae, Eresidae and Theridiidae). *American Zoologist* 12:419–426.

Laing, D. J. 1979. Studies on populations of the tunnel web spider *Porrhothele antipodiana* (Mygalomorphae: Dipluridae). Part II: Relationship with hunting wasps (Pompilidae). *Tautara* 24:1–20.

Lamore, D. H. 1958. The jumping spider, *Phidippus audax* Hentz, and the spider *Conopistha trigona* Hentz as predators of the basilica spider, *Allepeira lemniscata* Walckenaer in Maryland. *Proceedings of the Entomological Society of Washington* 60:286.

Larcher, S. F. & D. H. Wise. 1985. Experimental studies of the interactions between a web-invading spider and two host species. *Journal of Arachnology* 13:43–59.

Leclerc, J. & P. Blandin. 1990. Patch size, fine-scale co-occurrence and competition in forest litter linyphiids. *Acta Zoologica Fennica* 190:239–242.

Levine, S. H. 1976. Competitive interactions in ecosystems. *American Naturalist* 110:903–910.

Levins, R. 1968. *Evolution in Changing Environments.* Princeton, Princeton University Press.

Levins, R. 1975. Evolution in communities near equilibrium. In: *Ecology and Evolution of Communities,* M. L. Cody & J. M. Diamond, eds., pp. 16–50. Cambridge, Mass., Harvard University Press.

Levins, R. & R. Lewontin. 1980. Dialectics and reductionism in ecology. *Synthese* 43:47–78.

Louda, S. M. 1982. Inflorescence spiders: a cost/benefit analysis for the host plant, *Haplopappus venutus* Blake (Asteraceae). *Oecologia* 55:185–191.

Loughton, B. G., C. Derry & A. S. West. 1963. Spiders of the spruce budworm. *Memoirs of the Entomological Society of Canada* 31:249–268 (R. F. Morris, ed., *The Dynamics of Epidemic Spruce Budworm Populations*).

Lowrie, D. C. 1948. The ecological succession of spiders of the Chicago area dunes. *Ecology* 29:334–351.

Lowrie, D. C. 1987. Effects of diet on the development of *Loxosceles laeta* (Nicolet) (Araneae, Loxoscelidae). *Journal of Arachnology* 15:303–308.

Lubin, Y. 1974. Adaptive advantages and the evolution of colony formation in *Cyrtophora* (Araneae: Araneidae). *Zoological Journal of the Linnean Society* 54:321–329.

Łuczak, J. 1963. Differences in the structure of communities of web spiders in one type of environment (young pine forest). *Ekologia Polska–Seria A* 11:160–221.

Łuczak, J. 1966. The distribution of wandering spiders in different layers of the environment as a result of interspecies competition. *Ekologia Polska–Seria A* 14:234–244.

Łuczak, J. 1979. Spiders in agrocoenoses. *Polish Ecological Studies* 5:151–200.

MacArthur, R. H. & R. Levins. 1967. The limiting similarity, convergence and divergence of co-existing species. *American Naturalist* 101:377–385.

MacKay, W. P. 1982. The effect of predation of western widow spiders (Araneae: Theridiidae) on harvester ants (Hymenoptera: Formicidae). *Oecologia* 53:406–411.

Maelfait, J. P., J. Baert, J. Huble & A. de Kimpe. 1980. Life cycle timing, microhabitat preference and coexistence of spiders. In: *Proceedings of the Eighth International Congress of Arachnology*, pp. 69–73. Vienna, H. Egermann.

Manley, G. V., J. W. Butcher & M. Zarik. 1976. DDT transfer and metabolism in a forest litter macro-arthropod food chain. *Pedobiologia* 16:81–98.

Mansour, F. 1987. Spiders in sprayed and unsprayed cotton fields in Israel, their interactions with cotton pests and their importance as predators of the Egyptian cotton leaf worm, *Spodoptera littoralis*. *Phytoparasitica* 15:31–41.

Mansour, F. & W. H. Whitcomb. 1986. The spiders of a citrus grove in Israel and their role as biocontrol agents of *Ceroplastes floridensis* (Homoptera: Coccidae). *Entomophaga* 31:269–276.

Mansour, F., D. B. Richman & W. H. Whitcomb. 1983. Spider management in agroecosystems: habitat manipulation. *Environmental Management* 7:43–49.

Mansour, F., D. Rosen & A. Shulov. 1980a. Functional response of the spider *Chiracanthium mildei* (Arachnida: Clubionidae) to prey density. *Entomophaga* 25:313–316.

Mansour, F., D. Rosen, A. Shulov & H. N. Plaut. 1980b. Evaluation of spiders as biological control agents of *Spodoptera littoralis* larvae on apple in Israel. *Acta Oecologica/Oecologia Applicata* 1:225–232.

Markezich, A. L. 1987. Late season physiological adaptations of two syntopic araneid spiders. Unpubl. Ph.D. Dissertation, Illinois State University.

Martyniuk, J. 1983. Aspects of habitat choice and fitness in *Prolinyphia marginata* (Araneae: Linyphiidae): web-site selection, foraging dynamics, sperm competition and overwintering survival. Unpubl. Ph.D. Thesis, Sate University of New York at Binghamton.

Martyniuk, J. & D. H. Wise. 1985. Stage-biased overwintering survival of the filmy dome spider (Araneae, Linyphiidae). *Journal of Arachnology* 13:321–329.

Mather, M. H. & B. D. Roitberg. 1987. A sheep in wolf's clothing: tephritid flies mimic spider predators. *Science* 236:308–310.

McClay, C. L. & T. L. Hayward. 1987. Reproductive biology of the intertidal spider *Desis marina* (Araneae: Desidae) on a New Zealand rocky shore. *Journal of Zoology (London)* 211:357–372.

McIntosh, R. P. 1987. Pluralism in ecology. *Annual Review of Ecology and Systematics* 18:321–341.

McQueen, D. J. 1978. Field studies of growth, reproduction, and mortality in the burrowing wolf spider *Geolycosa domifex* (Hancock). *Canadian Journal of Zoology* 56:2037–2049.

McQueen, D. J. 1979. Interactions between the pompilid wasp *Anoplius relativus* (Fox) and the burrowing wolf spider *Geolycosa domifex* (Hancock). *Canadian Journal of Zoology* 57:542–550.

McQueen, D. J. & C. L. McClay. 1983. How does the intertidal spider *Desis marina* (Hector) remain under water for such a long time? *New Zealand Journal of Zoology* 10:383–392.

McReynolds, C. N. & G. A. Polis. 1987. Ecomorphological factors influencing prey use by two sympatric species of orb-web spiders, *Argiope aurantia* and *Argiope trifasciata* (Araneidae). *Journal of Arachnology* 15:371–384.

Miller, G. L. 1984. The influence of microhabitat and prey availability on burrow establishment of young *Geolycosa turricola* (Treat) and *G. micopany* Wallace (Araneae, Lycosidae): a laboratory study. *Psyche* 91:123–132.

Miyashita, K. 1968. Growth and development of *Lycosa T-insignata* Boes. et Str. (Araneae: Lycosidae) under different feeding conditions. *Applied Entomology and Zoology* 3:81–88.

Miyashita, K. 1969. Effects of locomotory activity, temperature and hunger on the respiratory rate of *Lycosa T-insignita* Boes. et Str. (Araneae: Lycosidae). *Applied Entomology and Zoology* 4:105–113.

Miyashita, T. 1986. Growth, egg production, and population density of the spider, *Nephila clavata* in relation to food conditions in the field. *Researches on Population Ecology* 28:135–149.

Morse, D. H. 1981. Prey capture by the crab spider *Misumena vatia* (Clerck) (Thomisidae) on three common native flowers. *American Midland Naturalist* 105:358–367.

Morse, D. H. 1986. Foraging decisions of crab spiders (*Misumena vatia*) hunting on flowers of different quality. *American Midland Naturalist* 116:341–347.

Morse, D. H. & R. S. Fritz. 1982. Experimental and observational studies of patch choice at different scales by the crab spider *Misumena vatia*. *Ecology* 63:172–182.

Morse, D. H. & R. S. Fritz. 1987. The consequences of foraging for reproductive success. In: *Foraging Behavior*, A. C. Kamil, J. R. Krebs & H. R. Pulliam, eds., pp. 443–455, New York, Plenum Press.

Motokawa, T. 1989. Sushi science and hamburger science. *Perspectives in Biology and Medicine* 32:489–504.

Moulder, B. C. & D. E. Reichle. 1972. Significance of spider predation in the energy dynamics of forest-floor arthropod communities. *Ecological Monographs* 42:473–498.

Mowat, F. 1975. *People of the Deer*. Toronto, McClelland and Stewart Limited.

Muma, M. H. & K. E. Muma. 1949. Studies on a population of prairie spiders. *Ecology* 30:485–503.

Murdoch, W. W. 1977. Stabilizing effects of spatial heterogeneity in predator–prey systems. *Theoretical Population Biology* 11:252–273.

Nakamura, K. 1972. The ingestion in wolf spiders. II. The expression of degree of hunger and amount of ingestion in relation to spider's hunger. *Researches on Population Ecology* 14:82–96.

Nakamura, K. 1977. A model for the functional response of a predator to varying prey densities, based on the feeding ecology of wolf spiders. *Bulletin of the National Institute of Agricultural Science of Japan, Series C* 31:28–89.

Nakamura, K. 1987. Hunger and starvation. In: *Ecophysiology of Spiders*, W. Nentwig, ed., pp. 287–295. Berlin, New York, London, Paris and Tokyo, Springer.

Nakasuji, F., K. Yamanaka & K. Kiritani. 1973. The disturbing effect of micryphantid spiders on the larval aggregations of the tobacco cutworm *Spodoptera litura*. *Kontyu* 41:220–227.

Nentwig, W. 1980. The selective prey of linyphiid-like spiders and of their space

304 · References

webs. *Oecologia* 45:236–243.

Nentwig, W. 1982. Epigeic spiders, their potential prey and competitors: relationship between size and frequency. *Oecologia* 55:130–136.

Nentwig, W. 1983. The prey of web-building spiders compared with feeding experiments (Araneae: Araneidae, Linyphiidae, Pholcidae, Agelenidae). *Oecologia* 56:132–139.

Nentwig, W. 1985a. Feeding ecology of the tropical spitting spider *Scytodes longipes* (Araneae, Scytodidae). *Oecologia* 65:284–288.

Nentwig, W. 1985b. Prey analysis of four species of tropical orb-weaving spiders (Araneae: Araneidae) and a comparison with araneids of the temperate zone. *Oecologia* 66:580–594.

Nentwig, W. (ed.). 1987. *Ecophysiology of Spiders.* Berlin, New York, London, Paris and Tokyo, Springer.

Nielsen, E. 1932. *The Biology of Spiders: With Especial Reference to the Danish Fauna.* Copenhagen, Lewin and Munksgaard. Vol. 1 in English, Vol. 2 in Danish.

Norberg, R. A. 1977. An ecological theory on foraging time and energetics and choice of optimal food-searching method. *Journal of Animal Ecology* 46:511–529.

Norberg, R. A. 1978. Energy content of some spiders and insects on branches of spruce (*Picea abies*) in winter: prey of certain passerine birds. *Oikos* 31:222–229.

Nørgaard, E. 1951. On the ecology of two lycosid spiders (*Pirata piraticus* and *Lycosa pullata*) from a Danish *Sphagnum* bog. *Oikos* 3:1–21.

Norton, R. A. 1973. Ecology of soil and litter spiders. In: *Proceedings of the First Soil Microcommunities Conference*, D. L. Dindal, ed., pp. 138–156. Springfield, Virginia (USA), National Technical Information Service, US Department of Commerce.

Nyffeler, M. 1982. Die ökologische Bedeutung der Spinnen in Forst-Ökosystemen, eine Literaturzusammenstellung. *Anzeiger für Schädlingskunde, Pflanzenschutz, Umweltschutz* 55:134–137.

Nyffeler, M. & G. Benz. 1978. Die Beutespektren der Netzspinnen *Argiope bruennichi* (Scop.), *Araneus quadratus* Cl. und *Agelena labyrinthica* (Cl.) in Ödlandwiesen bei Zürich. *Revue suisse de Zoologie* 85:747–757.

Nyffeler, M. & G. Benz. 1979a. Nischenüberlappung bezüglich der Raumund Nahrungskomponenten bei Krabbenspinnen (Araneae: Thomisidae) und Wolfspinnen (Araneae: Lycosidae) in Mähwiesen. *Revue suisse de Zoologie* 86:855–865.

Nyffeler, M. & G. Benz. 1979b. Zur ökologischen Bedeutung der Spinnen der Vegetationsschicht von Getreide- und Rapsfeldern bei Zürich (Schweiz). *Zeitschrift für angewandte Entomologie* 87:348–376.

Nyffeler, M. & G. Benz. 1981. Ökologische Bedeutung der Spinnen als Insektenprädotoren in Wiesen und Getreidefeldern. *Mitteilungen der deutschen Gesellschaft für allgemeine und angewandte Entomologie* 3:33–35.

Nyffeler, M. & G. Benz. 1987. Spiders in natural pest control: a review. *Zeitschrift für angewandte Entomologie* 103:321–339.

Nyffeler, M. & G. Benz. 1988. Prey and predatory importance of micryphantid spiders in winter wheat fields and hay meadows. *Zeitschrift für angewandte Entomologie* 104:190–197.

Nyffeler, M. & R. G. Breene. 1990. Evidence of low daily food consumption by wolf spiders in meadowland and comparison with other cursorial hunters. *Zeitschrift für angewandte Entomologie* 110:73–81.

Nyffeler, M., D. A. Dean & W. L. Sterling. 1987a. Evaluation of the importance of the striped lynx spider, *Oxyopes salticus* (Araneae: Oxyopidae), as a predator in Texas cotton. *Environmental Entomology* 16:1114–1123.

Nyffeler, M., D. A. Dean & W. L. Sterling. 1987b. Predation by green lynx spider, *Peucetia viridans* (Araneae: Oxyopidae), inhabiting cotton and woolly croton plants in East Texas. *Environmental Entomology* 16:355–359.

Nyffeler, M., C. D. Dondale & J. H. Redner. 1986. Evidence for displacement of a North American spider, *Steatoda borealis* (Hentz), by the European species *S. bipunctata* (Linnaeus) (Araneae: Theridiidae). *Canadian Journal of Zoology* 64:867–874.

Olechowicz, E. 1971. Productivity investigation of two types of meadows in the Vistula Valley. VIII. The number of emerged Diptera and their elimination. *Ekologia Polska* 19:183–195.

Olive, C. W. 1981. Optimal phenology and body-size of orb-weaving spiders: foraging constraints. *Oecologia* 49:83–87.

Olive, C. W. 1982. Behavioral responses of a sit-and-wait predator to spatial variation in foraging gain. *Ecology* 63:912–920.

Oraze, M. J. & A. A. Grigarick. 1989. Biological control of aster leafhopper (Homoptera: Cicadellidae) and midges (Diptera: Chironomidae) by *Pardosa ramulosa* (Araneae: Lycosidae) in California rice fields. *Journal of Economic Entomology* 82:745–749.

Otto, C. & B. S. Svensson. 1982. Structure of communities of ground-living spiders along altitudinal gradients. *Holarctic Ecology* 5: 35–47.

Pacala, S. & J. Roughgarden. 1984. Control of arthropod abundance by *Anolis* lizards on St. Eustatius (Neth. Antilles). *Oecologia* 64:160–162.

Pasquet, A. 1984a. Predatory site selection and adaptation of the trap in four species of orb-weaving spiders. *Biology of Behaviour* 9:3–20.

Pasquet, A. 1984b. Répartition de deux expèces d'araignées orbitèles, *Araneus Marmoreus* (clerk) et *A. Diadematus* (clerk), dans une prairie en friches. *Biology of Behaviour* 9:321–331.

Pasquet, A. 1984c. Proies capturées et stratégies prédatrices chez deux espèces d'araignées orbitèles: *Argiope bruennichi* et *Araneus marmoreus*. *Entomologia experimentalis et applicata* 36:177–184.

Pętal, J. 1978. The role of ants in ecosystems. In: *Production Ecology of Ants and Termites*, M. V. Brian, ed., pp. 293–325. Cambridge, New York and Melbourne, Cambridge University Press.

Pętal, J. & A. Breymeyer. 1969. Reduction of wandering spiders by ants in a *Stellario–Deschampsietum* meadow. *Bulletin de L'Académie Polonaise des Sciences, Cl. II, Série des sciences biologiques* 17:239–244.

Pianka, E. R. 1973. The structure of lizard communities. *Annual Review of Ecology and Systematics* 4:53–74.

Pianka, E. R. 1983. *Evolutionary Ecology*. New York, Harper and Row.

Poinar, G. O. Jr. 1987. Nematode parasites of spiders. In: *Ecophysiology of Spiders*, W. Nentwig, ed., pp. 299–308. Berlin, New York, London, Paris and Tokyo, Springer.

Pointing, P. J. 1965. Some factors influencing the orientation of the spider *Frontinella communis* (Hentz) in its web (Araneae: Linyphiidae). *Canadian Entomologist*

97:69–78.

Polis, G. A. & S. J. McCormick. 1986. Scorpions, spiders and solpugids: predation and competition among distantly related taxa. *Oecologia* 71:111–116.

Polis, G. A. & S. J. McCormick. 1987. Intraguild predation and competition among desert scorpions. *Ecology* 68:332–343.

Polis, G. A., C. A. Myers & R. D. Holt. 1989. The ecology and evolution of intraguild predation: potential competitors that eat each other. *Annual Review of Ecology and Systematics* 20:297–330.

Post, W. M. & S. E. Riechert. 1977. Initial investigation into the structure of spider communities. *Journal of Animal Ecology* 46:729–749.

Provencher, L. & D. Coderre. 1987. Functional responses and switching of *Tetragnatha laboriosa* Hentz (Araneae: Tetragnathidae) and *Clubiona pikei* Gertsch (Araneae: Clubionidae) for the aphids *Rhopalosiphum maidis* (Fitch) and *Rhopalosiphum padi* (L.) (Homoptera: Aphididae). *Environmental Entomology* 16:1305–1309.

Rehnberg, B. G. 1987. Selection of spider prey by *Trypoxylon politum* (Say) (Hymenoptera: Sphecidae). *Canadian Entomologist* 119:189–194.

Reichle, D. E. & D. A. Crossley. 1965. Radiocesium dispersion in a cryptozoan food web. *Health Physics* 11:1375–1384.

Reynoldson, T. B. & L. S. Bellamy. 1971. The establishment of interspecific competition in field populations, with an example of competition in action between *Polycelis nigra* (Mull.) and *P. tenuis* (Ijima) (Turbellaria, Tricladida). In: *Proceedings of the Advanced Study Institute on the Dynamics of Numbers of Populations*, pp. 282–297. Oosterbeck (1970).

Rice, W. R. 1989. Analyzing tables of statistical tests. *Evolution* 43:223–225.

Richter, C. J. J. 1970. Aerial dispersal in relation to habitat in eight wolf spider species (*Pardosa*, Araneae, Lycosidae). *Oecologia* 5:200–214.

Ricklefs, R. E. 1983. *The Economy of Nature*. New York, Chiron Press. Copyrighted 1990 by W. H. Freeman, New York.

Riechert, S. E. 1974a. Thoughts on the ecological significance of spiders. *Bioscience* 24:352–356.

Riechert, S. E. 1974b. The pattern of local web distribution in a desert spider: mechanisms and seasonal variation. *Journal of Animal Ecology* 43:733–746.

Riechert, S. E. 1976. Web-site selection in the desert spider *Agelenopsis aperta*. *Oikos* 27:311–315.

Riechert, S. E. 1978a. Energy-based territoriality in populations of the desert spider *Agelenopsis aperta* (Gertsch). *Symposia of the Zoological Society of London* 42:211–222.

Riechert, S. E. 1978b. Games spiders play: behavioral variability in territorial disputes. *Behavioral Ecology and Sociobiology* 3:135–162.

Riechert, S. E. 1979. Games spiders play. II. Resource assessment strategies. *Behavioral Ecology and Sociobiology* 6:121–128.

Riechert, S. E. 1981. The consequences of being territorial: spiders, a case study. *American Naturalist* 117:871–892.

Riechert, S. E. 1982. Spider interaction strategies: communication vs. coercion. In: *Spider Communication: Mechanisms and Ecological Significance*, P. N. Witt & J. S. Rovner, eds., pp. 281–315. Princeton, Princeton University Press.

Riechert, S. E. 1986. Spider fights as a test of evolutionary game theory. *American Scientist* 74:604–610.

Riechert, S. E. 1988. The energetic costs of fighting. *American Zoologist* 28:877–884.

Riechert, S. E. & L. Bishop. 1990. Prey control by an assemblage of generalist predators: spiders in garden test systems. *Ecology* 71:1441–1450.

Riechert, S. E. & A. B. Cady. 1983. Patterns of resource use and tests for competitive release in a spider community. *Ecology* 64:899–913.

Riechert, S. E. & R. G. Gillespie. 1986. Habitat choice and utilization in web-building spiders. In: *Spiders – Webs, Behavior, and Evolution*, W. A. Shear, ed., pp. 23–48. Stanford, Calif., Stanford University Press.

Riechert, S. E. & J. Harp. 1987. Nutritional ecology of spiders. In: *Nutritional Ecology of Insects, Mites, and Spiders*, F. Slansky, Jr. & J. G. Rodriguez, eds., pp. 645–672. New York, John Wiley & Sons.

Riechert, S. E. & A. V. Hedrick. 1990. Levels of predation and genetically based anti-predator behaviour in the spider, *Agelenopsis aperta*. *Animal Behaviour* 40:679–687.

Riechert, S. E. & T. Lockley. 1984. Spiders as biological control agents. *Annual Review of Entomology* 29:299–320.

Riechert, S. E. & J. Łuczak. 1982. Spider foraging: behavioral responses to prey. In: *Spider Communication: Mechanisms and Ecological Significance*, P. N. Witt & J. S. Rovner, eds., pp. 353–385. Princeton, Princeton University Press.

Riechert, S. E. & C. R. Tracy. 1975. Thermal balance and prey availability: bases for a model relating web-site characteristics to spider reproductive success. *Ecology* 56:265–285.

Riechert S. E., W. G. Reeder & T. A. Allen. 1973. Patterns of spider distribution [(*Agelenopsis aperta* (Gertsch)] in desert grassland and recent lava bed habitats, south-central New Mexico. *Journal of Animal Ecology* 42:19–35.

Robinson, J. V. 1981. The effect of architectural variation in habitat on a spider community: an experimental field study. *Ecology* 62:73–80.

Robinson, M. H. & B. Robinson. 1973. Ecology and behavior of the giant wood spider *Nephila maculata* (Fabr.) in New Guinea. *Smithsonian Contribution to Zoology* 149:1–76.

Robinson, M. H. & B. C. Robinson. 1978. Thermoregulation in orb-web spiders: new descriptions of thermoregulatory postures and experiments on the effects of posture and coloration. *Zoological Journal of the Linnean Society* 64:87–102.

Robinson, M. H., B. C. Robinson, F. M. Murphy & W. S. Corley. 1986. Egg-sac burying behavior in *Nephila maculata*: a specialized adaptation. In: *Proceedings of the Ninth International Congress of Arachnology*, W. G. Eberhard, Y. D. Lubin & B. C. Robinson, eds., pp. 245–254. Washington, DC, Smithsonian Institution Press.

Roeloffs, R. & S. E. Riechert. 1988. Dispersal and population-genetic structure of the cooperative spider, *Agelena consociata*, in West African rainforest. *Evolution* 42:173–183.

Rotenberry, J. T. & J. A. Wiens. 1985. Statistical power analysis and community-wide patterns. *American Naturalist* 125:164–168.

Rovner, J. S. 1987. Nests of terrestrial spiders maintain a physical gill: floods and evolution of silk construction. *Journal of Arachnology* 14:327–338.

Rubenstein, D. I. 1987. Alternative reproductive tactics in the spider *Meta segmentata*. *Behavioral Ecology and Sociobiology* 20:229–237.

Rushton, S. P., C. J. Topping & M. D. Eyre. 1987. The habitat preferences of grassland spiders as identified using Detrended Correspondence Analysis

(DECORANA). *Bulletin of the British Arachnological Society* 7:165–170.

Rypstra, A. L. 1979. Foraging flocks of spiders: a study of aggregate behavior in *Cyrtophora citricola* Forskal (Araneae: Araneidae) in West Africa. *Behavioral Ecology and Sociobiology* 5:291–300.

Rypstra, A. L. 1981. The effect of kleptoparasitism on prey consumption and web relocation in a Peruvian population of the spider *Nephila clavipes*. *Oikos* 37.179–182.

Rypstra, A. L. 1983. The importance of food and space in limiting web-spider densities; a test using field enclosures. *Oecologia* 59:312–316.

Rypstra, A. L. 1984. A relative measure of predation on web-spiders in temperate and tropical forests. *Oikos* 43:129–132.

Rypstra, A. L. 1985. Aggregations of *Nephila clavipes* (L.) (Araneae, Araneidae) in relation to prey availability. *Journal of Arachnology* 13:71–78.

Rypstra, A. L. 1986. Web spiders in temperate and tropical forests: relative abundance and environmental correlates. *American Midland Naturalist* 115:42–51.

Rypstra, A. L. 1989. Foraging success of solitary and aggregated spiders: insights into flock formation. *Animal Behaviour* 37:274–281.

Sasaba, T. & K. Kiritani. 1972. Evaluation of mortality factors with special reference to parasitism of the green rice leafhopper, *Nephotettix cincticeps*. *Applied Entomology and Zoology* 7:83–93 (cited by Kiritani & Kakiya 1975).

Sasaba, T., K. Kiritani & T. Urabe. 1973. A preliminary model to simulate the effect of insecticides on a spider–leafhopper system in the paddy field. *Researches on Population Ecology* 15:9–22 (cited by Kiritani & Kakiya 1975).

Savory, T.H. 1964. *Arachnida*. London, Academic Press.

Schaefer, M. 1972. Ökologische Isolation und die Bedeutung des Konkurrenzfaktors am Beispiel des Verteilungsmusters der Lycosiden einer Küstenlandschaft. *Oecologia* 9:171–202.

Schaefer, M. 1974. Experimentelle Untersuchungen zur Bedeutung der interspezifischen Konkurrenz bei 3 Wolfspinnen-Arten (Araneida: Lycosidae) einer Salzwiese. *Zoologische Jahrbücher, Abteilung Systematik, Ökologie und Geographie der Tiere* 101:213–235.

Schaefer, M. 1975. Experimental studies on the importance of interspecies competition for the lycosid spiders in a salt marsh. In: *Proceedings of the Sixth International Arachnological Congress*, pp. 86–90. Amsterdam.

Schaefer, M. 1977. Zur Bedeutung des Winters für die Populationsdynamik von vier Spinnenarten (Araneida). *Zoologischer Anzeiger* 199:77–88.

Schaefer, M. 1978. Some experiments on regulation of population density in the spider *Floronia bucculenta* (Araneidae: Linyphiidae). *Symposia of the Zoological Society of London* 42:203–210.

Schaefer, M. 1981. Untersuchungen über Räuber-Beute-Systeme bei Arthropoden einiger Lebensgemeinschaften der offenen Landschaft. *Jahresberichte des naturwissenschaftlichen Vereins Wuppertal*. 34:48–53.

Schaefer, M. 1988. Struktur und Funktion von saprophagen terrestrischen Bodentiergruppen in einem Kalkbuchenwald. *Arbeitsbericht zum Forschungsvorhaben Scha 352/1–2 (Göttingen)*, 85 pp. (Unpublished research report to the 'Deutsche Forschungsgemeinschaft'.)

Schauermann, J. 1982. Verbesserte Extraktion der terrestrischen Bodenfauna im

Vielfachgerät modifiziert nach Kempson und Macfadyen. *Mitteilungen aus dem Sonderforschungsbereich (Ökosysteme auf Kalkestein) 135*(1):47–50.

Scheidler, M. 1989. Niche partitioning and density distribution in two species of *Theridion* (Theridiidae, Araneae) on thistles. *Zoologischer Anzeiger* 223:49–56.

Schoener, T. W. 1968. The *Anolis* lizards of Bimini: resource partitioning in a complex fauna. *Ecology* 49:704–726.

Schoener, T. W. 1974. Some methods for calculating competition coefficients from resource-utilization spectra. *American Naturalist* 108:322–340.

Schoener, T. W. 1982. The controversy over interspecific competition. *American Scientist* 70:586–595.

Schoener, T. W. 1983a. Field experiments on interspecific competition. *American Naturalist* 122:240–285.

Schoener, T. W. 1983b. Interspecific competition – response to a letter to the editor from J. A. Wiens. *American Scientist* 71:235.

Schoener, T. W. 1986. Patterns in terrestrial vertebrate versus arthropod communities: do systematic differences in regularity exist? In: *Community Ecology*, J. Diamond & T. J. Case, eds., pp. 556–586. New York, Harper and Row.

Schoener, T. W. 1987. A brief history of optimal foraging ecology. In: *Foraging Behavior*, A. C. Kamil, J. Krebs & H. R. Pulliam, eds., pp. 5–67. New York, Plenum Publishing Corporation.

Schoener, T. W. 1988. Leaf damage in island buttonwood, *Conocarpus erectus*: correlations with pubescence, island area, isolation and the distribution of major carnivores. *Oikos* 53:252–266.

Schoener, T. W. 1989. Food webs from the small to the large. *Ecology* 70:1559–1589.

Schoener, T. W. & D. A. Spiller. 1987. Effect of lizards on spider populations: manipulative reconstruction of a natural experiment. *Science* 236:949–952.

Schoener, T. W. & C. A. Toft. 1983a. Spider populations: extraordinarily high densities on islands without top predators. *Science*: 219:1353–1355.

Schoener, T. W. & C. A. Toft. 1983b. Dispersion of a small-island population of the spider *Metepeira datona* (Araneae: Araneidae) in relation to web-site availability. *Behavioral Ecology and Sociobiology* 12:121–128.

Shear, W. A. 1970. The evolution of social phenomena in spiders. *Bulletin of the British Arachnological Society* 1:65–76.

Shear, W. A. (ed.) 1986. *Spiders – Webs, Behavior, and Evolution*. Stanford, Calif., Stanford University Press.

Shelly, T. E. 1984. Prey selection by the neotropical spider *Micrathena schreibersi* with notes on web-site tenacity. *Proceedings of the Entomological Society of Washington* 86:493–502.

Sih, A., P. Crowley, M. McPeek, J. Petranka & K. Strohmeier. 1985. Predation, competition, and prey communities: a review of field experiments. *Annual Review of Ecology and Systematics* 16:269–311.

Simberloff, D. 1982. The status of competition theory in ecology. *Annales Zoologici Fennici* 19:241–253.

Smith, R. B. & T. P. Mommsen. 1984. Pollen feeding in an orb-weaving spider. *Science* 226:1330–1333.

Smith, R. B. & W. G. Wellington. 1986. The functional response of a juvenile orb-weaving spider. In: *Proceedings of the Ninth International Congress of Arachnology*,

W. G. Eberhard, Y. D. Lubin & B. C. Robinson, eds., pp. 275–279. Washington, DC, Smithsonian Institution Press.

Smith-Trail, D. 1980. Predation by *Argyrodes* (Theridiidae) on solitary and communal spiders. *Psyche* 87:349–355.

Sokal, R. R. & J. Rohlf. 1987. *Introduction to Biostatistics.* New York, W. H. Freeman.

Solomon, M. E. 1949. The natural control of animal populations. *Journal of Animal Ecology* 18.1–35.

Spiller, D. A. 1984a. Competition between two spider species: experimental field study. *Ecology* 65:909–919.

Spiller, D. A. 1984b. Seasonal reversal of competitive advantage between two spider species. *Oecologia* 64:322–331.

Spiller, D. A. 1986a. Consumptive-competition coefficients: an experimental analysis with spiders. *American Naturalist* 127:604–614.

Spiller, D. A. 1986b. Interspecific competition between spiders and its relevance to biological control by general predators. *Environmental Entomology* 15:177–181.

Spiller, D. A. & T. W. Schoener. 1988. An experimental study of the effect of lizards on web–spider communities. *Ecological Monographs* 58:51–77.

Spiller, D. A. & T. W. Schoener. 1990a. Lizards reduce food consumption by spiders: mechanisms and consequences. *Oecologia* 83:150–161.

Spiller, D. A. & T. W. Schoener. 1990b. A terrestrial field experiment showing the impact of eliminating top predators on foliage damage. *Nature* 347:469–472.

Stevenson, B. G. & D. L. Dindal. 1982. Effect of leaf shape on forest litter spiders: community organization and microhabitat selection of immature *Enoplognatha ovata* (Clerck)(Theridiidae). *Journal of Arachnology* 10:165–178.

Stippich, G. 1989. Die Bedeutung von natürlichen und künstlichen Strukturelementen für die Besiedelung des Waldbodens durch Spinnen (Zur Funktion der Fauna in einem Mullbuchenwald 14). *Verhandlungen der Gesellschaft für Ökologie* 17:293–298.

Stone, M. 1979. *Ancient Mirrors of Womanhood.* New York, New Sibylline Books. Republished 1984 by Beacon Press, Boston.

Stowe, M. K. 1986. Prey specialization in the Araneidae. In: *Spiders – Webs, Behavior, and Evolution*, W. A. Shear, ed., pp. 101–131. Stanford, Calif., Stanford University Press.

Stratton, G. E. 1984. Behavioral studies of wolf spiders: a review of recent studies. *Revue arachnologique* 6:57–70.

Strong, D. R. 1980. Null hypotheses in ecology. *Synthese* 43:271–285.

Strong, D. R. 1984. Density-vague ecology and liberal population regulation in insects. In: *A New Ecology – Novel Approaches to Interactive Systems*, P. W. Price, C. N. Slobodchikoff & W. S. Gaud, eds., pp. 313–327. New York, John Wiley & Sons.

Strong, D. R., D. Simberloff, L. G. Abele & A. B. Thistle (eds.). 1984. *Ecological Communities: Conceptual Issues and the Evidence.* Princeton, Princeton University Press.

Sunderland, K. D. 1988. Quantitative methods for detecting invertebrate predation occurring in the field. *Annals of Applied Biology* 112:201–224.

Sunderland, K. D., N. E. Crook, D. L. Stacey & B. J. Fuller. 1987. A study of feeding by polyphagous predators on cereal aphids using ELISA and gut dissection. *Journal of Applied Ecology* 24:907–933.

Suter, R. B. 1981. Behavioral thermoregulation: solar orientation in *Frontinella communis* (Linyphiidae), a 6-mg spider. *Behavioral Ecology and Sociobiology* 8:77–81.

Suwa, M. 1986. Space partitioning among the wolf spider *Pardosa amentata* species group in Hokkaido, Japan. *Researches on Population Ecology* 28:231–252.

Syrek, D. & B. Janusz. 1977. Spatial structure of populations of spiders *Trochosa terricola* Thorell, 1856, and *Pardosa pullata* (Clerck, 1758). *Ekologia Polska* 25:107–113.

Tanaka, K. 1984. Rate of predation by a kleptoparasitic spider, *Argyrodes fissifrons*, upon a large host spider, *Agelena limbata*. *Journal of Arachnology* 12:363–367.

Tanaka, K. 1991. Food consumption and diet composition of the web-building spider *Agelena limbata* in two habitats. *Oecologia* 86:8–15.

Tanaka, K. & Y. Itô. 1982. Decrease in respiratory rate in a wolf spider, *Pardosa astrigera* (L. Koch), under starvation. *Researches on Population Ecology* 24:360–374.

Tanaka, K., Y. Itô & T. Saito. 1985. Reduced respiratory quotient by starvation in a wolf spider, *Pardosa astrigera* (L. Koch). *Comparative Biochemistry and Physiology* 80A:415–418.

Taub, M. L. 1977. Differences facilitating the coexistence of two sympatric, orb-web spiders, *Argiope aurantia* Lucas and *Argiope trifasciata* (Forskal) (Araneidae, Araneae). Unpubl. Master's Thesis, University of Maryland.

Toft, C. A. & T. W. Schoener. 1983. Abundance and diversity of orb spiders in 106 Bahamian islands: biogeography at an intermediate trophic level. *Oikos* 41:411–426.

Toft, C. A. & P. J. Shea. 1983. Detecting community-wide patterns: estimating power strengthens statistical inference. *American Naturalist* 122:618–625.

Toft, S. 1978. Phenology of some Danish beech-wood spiders. *Natura Jutlandica* 20:285–304.

Toft, S. 1980. Body size relations of sheet-web spiders in Danish *Calluna* heaths. In: *Proceedings of the Eighth International Congress of Arachnology*, pp. 161–164. Vienna, H. Egermann.

Toft, S. 1986. Field experiments on competition among web spiders. In: *Proceedings of the Ninth International Congress of Arachnology*, W. G. Eberhard, Y. D. Lubin & B. C. Robinson, eds., p. 328 (Abstract). Washington, DC, Smithsonian Institution Press.

Toft, S. 1987. Microhabitat identity of two species of sheet-web spiders: field experimental demonstration. *Oecologia* 72:216–220.

Toft, S. 1988. Interference by web take-over in sheet-web spiders. *XI. Europäisches Arachnologisches Colloquium, Technische Universität Berlin, Dokumentation Kongresse und Tagungen* 38:48–59.

Toft, S. 1990. Interactions among two coexisting *Linyphia* spiders. *Acta Zoologica Fennica* 190:367–372.

Tolbert, W. W. 1976. Population dynamics of the orb-weaving spiders *Argiope trifasciata* and *Argiope aurantia* (Araneae: Araneidae). Unpubl. Ph.D. Dissertation, University of Tennessee.

Traczyk, T., H. Traczyk & D. Pasternak. 1976. The influence of intensive mineral fertilization on the yield and floral composition of meadows. *Polish Ecological Studies* 2:36–47 (cited by Kajak 1981).

Trambarulo, A. 1981. Public attitudes towards cockroaches and their potential biological control by *Heteropoda venatoria* L. (Sparassidae). *American Arachnology* 24:25 (abstract).

Tretzel, E. 1955. Intragenerische Isolation und interspezifische Konkurrenz bei Spinnen. *Zeitschrift für Morphologie und Ökologie Tierre* 44:43–162.

Turnbull, A. L. 1960a. The spider population of a stand of oak (*Quercus robur* L.) in Wytham Wood, Berks., England. *Canadian Entomologist* 92:110–124.

Turnbull, A. L. 1960b. The prey of the spider *Linyphia triangularis* (Clerck) (Araneae, Linyphiidae). *Canadian Journal of Zoology* 35:859–873.

Turnbull, A. L. 1964. The search for prey by a web-building spider *Achaearanea tepidariorum* (C. L. Koch) (Araneae, Theridiidae). *Canadian Entomologist* 96:568–579.

Turnbull, A. L. 1973. Ecology of the true spiders (Araneomorphae). *Annual Review of Entomology* 18:305–348.

Turner, M. & G. A. Polis. 1979. Patterns of co-existence in a guild of raptorial spiders. *Journal of Animal Ecology* 48:509–520.

Uetz, G. W. 1974. A method for measuring habitat space in studies of hardwood forest litter arthropods. *Environmental Entomology* 3:313–315.

Uetz, G. W. 1975. Temporal and spatial variation in species diversity of wandering spiders (Araneae) in deciduous forest litter. *Environmental Entomology* 4:719–724.

Uetz, G. W. 1976. Gradient analysis of spider communities in a streamside forest. *Oecologia* 22:373–385.

Uetz, G. W. 1977. Coexistence in a guild of wandering spiders. *Journal of Animal Ecology* 46:531–542.

Uetz, G. W. 1979. The influence of variation in litter habitats on spider communities. *Oecologia* 40:29–42.

Uetz, G. W. 1985. Ecology and behavior of *Metepeira spinipes* (Araneae: Araneidae), a colonial web-building spider from Mexico. *National Geographic Research Reports* 19:597–609.

Uetz, G. W. (ed.). 1986a. Symposium: social behavior in spiders. *Journal of Arachnology* 14:145–281.

Uetz, G. W. 1986b. Web building and prey capture in communal orb weavers. In: *Spiders – Webs, Behavior, and Evolution*, W. A. Shear, ed., pp. 207–231. Stanford, Calif., Stanford University Press.

Uetz, G. W. 1988. Risk-sensitivity and foraging in colonial spiders. In: *Ecology of Social Behavior*, C. N. Slobodchikoff, ed., pp. 353–377. New York, Academic Press.

Uetz, G. W. 1989. The 'ricochet effect' and prey capture in colonial spiders. *Oecologia* 81:154–159.

Uetz, G. W. 1991. Habitat structure and spider foraging. In: *Habitat Structure: The Physical Arrangement of Objects in Space*, S. S. Bell, E. D. McCoy & H. R. Mushinsky, eds., pp. 325–348. London, Chapman and Hall.

Uetz, G. W. & J. M. Biere. 1980. Prey of *Micrathena gracilis* (Walckenaer) (Araneae: Araneidae) in comparison with artificial webs and other trapping devices. *Bulletin of the British Arachnological Society* 5:101–107.

Uetz, G. W. & G. Denterlein. 1979. Courtship behavior, habitat, and reproductive isolation in *Schizocosa rovneri* (Uetz and Dondale) (Araneae: Lycosidae). *Journal of Arachnology* 7:121–128.

Uetz, G. W., A. D. Johnson & D. W. Schemske. 1978. Web placement, web structure, and prey capture in orb-weaving spiders. *Bulletin of the British Arachnological Society* 4:141–148.

Uetz, G. W., K. L. Van Der Laan, G. F. Summers, P. A. K. Gibson & L. L. Getz. 1979. The effects of flooding on floodplain arthropod distribution, abundance and community structure. *American Midland Naturalist* 101:286–299.

Usher, M. B., R. G. Booth & K. E. Sparkes. 1982. A review of progress in understanding the organization of communities of soil arthropods. *Pedobiologia* 23:126–144.

Vandermeer, J., B. Hazlett & B. Rathcke. 1985. Indirect facilitation and mutualism. In: *The Biology of Mutualism*, D. H. Boucher, ed., pp. 326–343. Oxford and New York, Oxford University Press.

Van Dyke, D. & D. C. Lowrie. 1975. Comparative life histories of the wolf spiders *Pardosa ramulosa* and *P. sierra*. *Southwestern Naturalist* 20:29–44.

Van Hook, R. I., Jr. 1971. Energy and nutrient dynamics of spider and orthopteran populations in a grassland ecosystem. *Ecological Monographs* 41:1–26.

Vermeulen, C. & A. Kessler. 1980. Coexistence of webspiders (Linyphiidae, Argiopidae) in buckthorn bushes (*Hippophae rhamnoides* L.). In: *Proceedings of the Eighth International Congress of Arachnology*, pp. 155–159. Vienna, H. Egermann.

Vince, S. W., I. Valiela & J. M. Teal. 1981. An experimental study of the structure of herbivorous insect communities in a salt marsh. *Ecology* 62:1662–1678.

Vollrath, F. 1976. Konkurrenzvermeidung bei tropischen kleptoparasitischen Haubennetzspinnen der Gattung *Argyrodes* (Arachnida: Araneae: Theridiidae). *Entomologica Germanica* 3:104–108.

Vollrath, F. 1979. Behavior of the kleptoparasitic spider *Argyrodes elevatus* (Araneae: Theridiidae). *Animal Behaviour* 27:515–521.

Vollrath, F. 1982. Colony formation in a social spider. *Zeitschrift für Tierpsychologie* 60:313–324.

Vollrath, F. 1985. Web spider's dilemma: a risky move or site dependent growth. *Oecologia* 68:69–72.

Vollrath, F. 1987. Kleptobiosis in spiders. In: *Ecophysiology of Spiders*, W. Nentwig, ed., pp. 274–286. Berlin, New York, London, Paris and Tokyo, Springer.

Vollrath, F. 1988. Spider growth as an idicator of habitat quality. *Bulletin of the British Arachnological Society* 7:217–219.

Weigle, M. 1982. *Spiders and Spinsters*. Albuquerque, NM, University of New Mexico Press.

Wiehle, H. 1928. Beiträge zur Biologie der Araneen, insbesondere zur Kenntnis des Radnetzbaues. *Zeitschrift für Morphologie und Ökologie Tierre* 22:349–400.

Wiens, J. A. 1977. On competition and variable environments. *American Scientist* 65:590–597.

Wiens, J. A. 1983. Interspecific competition – a letter to the editor. *American Scientist* 71:234–235.

Wilson, E. O. 1978. *On Human Nature*. Cambridge, Mass., Harvard University Press.

Wingerden, W. K. R. van. 1975. Population dynamics of *Erigone arctica* (White) (Araneae, Linyphiidae). In: *Proceedings of the Sixth International Arachnological Congress*, pp. 71–76. Amsterdam.

Wingerden, W. K. R. van. 1978. Population dynamics of *Erigone arctica* (White) (Araneae, Linyphiidae) II. *Symposia of the Zoological Society of London* 42:195–202.

Wise, D. H. 1974. Role of food supply in the population dynamics of the spider *Linyphia marginata*. Unpubl. Ph.D. Dissertation, University of Michigan.

Wise, D. H. 1975. Food limitation of the spider *Linyphia marginata*: experimental field studies. *Ecology* 56:637–646.

Wise, D. H. 1976. Variable rates of maturation of the spider, *Neriene radiata* (*Linyphia marginata*). *American Midland Naturalist* 96:66–77.

Wise, D. H. 1979. Effects of an experimental increase in prey abundance upon the reproductive rates of two orb-weaving spider species (Araneae: Araneidae). *Oecologia* 41:289–300.

Wise, D. H. 1981. Inter- and intraspecific effects of density manipulations upon females of two orb-weaving spiders (Araneae: Araneidae). *Oecologia* 48:252–256.

Wise, D. H. 1982. Predation by a commensal spider, *Argyrodes trigonum*, upon its host: an experimental study. *Journal of Arachnology* 10:111–116.

Wise, D. H. 1983. Competitive mechanisms in a food-limited species: relative importance of interference and exploitative interactions among labyrinth spiders (Araneae: Araneidae). *Oecologia* 58:1–9.

Wise, D. H. 1984a. The role of competition in spider communities: insights from field experiments with a model organism. In: *Ecological Communities: Conceptual Issues and the Evidence*, D. R. Strong, D. Simberloff, L. G. Abele & A. B. Thistle, eds., pp. 42–53. Princeton, Princeton University Press.

Wise, D. H. 1984b. Phenology and life history of the filmy dome spider (Araneae: Linyphiide) in two local Maryland populations. *Psyche* 91:267–288.

Wise, D. H. 1987. Rearing studies with a spider exhibiting a variable phenology: no evidence of substantial genetic variation. *Bulletin of the British Arachnological Society* 7:107–110.

Wise, D. H. & J. L. Barata. 1983. Prey of two syntopic spiders with different web structures. *Journal of Arachnology* 11:271–281.

Wise, D. H. & P. R. Reillo. 1985. Frequencies of color morphs in four populations of *Enoplognatha ovata* (Araneae: Theridiidae) in eastern North America. *Psyche* 92:135–145.

Wise, D. H. & J. D. Wagner. 1992. Exploitative competition for prey among young stages of the wolf spider *Schizocosa ocreata*. *Oecologia* 91:7–13.

Witt, P. N., C. F. Reed & D. B. Peakall. 1968. *A Spider's Web: Problems in Regulatory Biology*. Berlin, New York, London, Paris and Tokyo, Springer.

Witt, P. N. & J. S. Rovner (eds.) 1982. *Spider Communication*. Princeton, Princeton University Press.

Wolters, V. 1985. Untersuchung zur Habitatbindung und Nahrungsbiologie der Springschwänze (Collembola) eines Laubwaldes unter besonderer Berücksichtigung ihrer Funktion in der Zersetzerkette. Unpubl. Ph.D. Dissertation, University of Göttingen.

Yamanaka, K., F. Nakasuji & K. Kiritani. 1973. Life tables of the tobacco cutworm *Spodoptera litura* and the evaluation of effectiveness of natural enemies (in Japanese; abstract in English). *Japanese Journal of Applied Entomology and Zoology* 16:205–214.

Yodzis, P. 1988. The indeterminancy of ecological interactions as perceived through perturbation experiments. *Ecology* 69:508–515.

Yoshida, M. 1981. Preliminary study on the ecology of three horizontal orb weavers, *Tetragnatha praedonia, T. japonica* and *T. pinicola* (Araneae: Tetragnathidae). *Acta Arachnologica* 30:49–64.

Young, O. P. & T. C. Lockley. 1985. The striped lynx spider, *Oxyopes salticus*, in agroecosystems. *Entomophaga* 30:329–346.

Zimmermann, M. & J. R. Spence. 1989. Prey use of the fishing spider *Dolomedes triton* (Pisauridae, Araneae): an important predator of the neuston community. *Oecologia* 80:187–194.

Name index

Subject index

abiotic factors, 48, 110–11, 113, 127, 129, 131, 135, 139, 140, 160, 182, 187, 191, 195, 217–8, 262–3, 273, 284
Achaearanea tepidariorum, 75, 216, 263
aerial insect plankton, 135
Agelena consociata, 128
Agelenidae, 8–9, 17, 195–6, 202; *see also* *Agelena, Agelenopsis, Coelotes*
Agelenopsis
  aperta, 23, 110–11, 123–4, 130, 133, 155, 156, 159, 190–1, 193, 284
  naevia, 9
agroecosystems, 43–4, 151–2, 160–79, 252, 260, 262, 268, 284, 286–9
  spiders as biocontrol agents in, 160–76; field experiments, 165–76; indirect evidence, 162–5
air pollution, 192
allopatry, 54, 56
allotopy, 62, 84
amaurobiids, 202
Ananse, 3
Andropogon, 184
ants, 91, 120, 146–7, 243
anyphaenids, 265
aphids, 155–6, 162, 171–3
Arachne, 2, 100
Araneae, overview, 1–12
Araneidae, 4, 17, 23, 26, 37, 39, 117, 144, 182; *see also* *Araneus, Argiope, Cyclosa, Eustala, Mangora, Mastophora, Mecynogea, Meta, Metepeira, Micrathena, Nephila*
Araneomorphae, 12
Araneus spp., 35
  cavaticus, 75–6
  cornutus, 83, 209
  diadematus, 157, 210–11
  marmoreus, 85, 132
  quadratus, 83, 246–7
  trifolium, 46

architectural environment, 181–220
Arctosa littoralis, 182
Argiope spp., 120
  argentata, 77, 125, 127
  aurantia, 25, 37, 46–7, 64–6, 85–8, 96–8, 132, 272
  bruennichi, 83, 85
  keyserling, 32
  trifasciata, 31, 32, 37, 46–7, 64–6, 85–8, 96–8, 132, 272
Argyrodes spp., 223
  caudatus, 52
  elevatus, 52–3
  fissifrans, 119
  trigonum, 119, 121
Argyroneta aquatica, 2
Argyronetidae, 2
Artemisia, 212
Athena, 2
avian tail, 265

ballooning, 128–9
barley, 171–2
Barro Colorado Island, 52, 118
basilica spider, *see Mecynogea lemniscata*
beech, 78, 143, 148, 183, 194, 196, 201, 203, 206, 243
beetles, *see* Coleoptera
biocontrol, 142, 152, 160–76, 179, 262, 264, 268, 287
biotic factors, 45, 129, 140, 160, 181–2, 218, 220–1, 224, 262–3
birds, 3, 15, 113, 120–4, 128, 140, 181, 192, 265
bowl-and-doily spider, *see Frontinella pyramitela*
broomsedge, 184

cages, 31, 122, 145–6, 166, 215–16, 218, 226, 246–8, 257, 260, 274, 276
California, 31, 47, 49, 68, 166, 185–7